KV-638-842

Leeds
street by street

with the National Grid

The maps in this Atlas are based upon the Ordnance Survey
Maps with the sanction of the Controller of H.M. Stationery Office,
with additions obtained from Local Authorities. The Ordnance Survey
is not responsible for the accuracy of the National Grid
on this Production.

The representation on this map of a Road, Track or Footpath
is no evidence of existence of a right of way.

©Geographia

Geographia Ltd
63 Fleet Street, London E.C.4
Telephone 01-353 2701/2

ISBN 0 09 202 1107

Titles in this series

Title	Inches to mile	Size	SBN
Birmingham and West Midlands with Coventry	4	8½″ x 5½″	09 202070 4
Bristol with Bath and Weston-super-Mare	4	8½″ x 5½″	09 202080 1
Edinburgh	4	8½″ x 5½″	09 202090 9
Glasgow	4	8½″ x 5½″	09 202100 X
Leeds with Pudsey and Horsforth	4	8½″ x 5½″	09 202110 7
Liverpool	4	8½″ x 5½″	09 202120 4
London	3	7″ x 4½″	09 202130 1
Manchester	4	8½″ x 5½″	09 202140 9
Sheffield & Rotherham	4	8½″ x 5½″	09 202150 6
Stoke on Trent and Newcastle under Lyme	4	8½″ x 5½″	09 202160 3
Tyneside (Newcastle upon Tyne, Gateshead, Tynemouth, South Shields) and Sunderland	5	8½″ x 5½″	09 202170 0

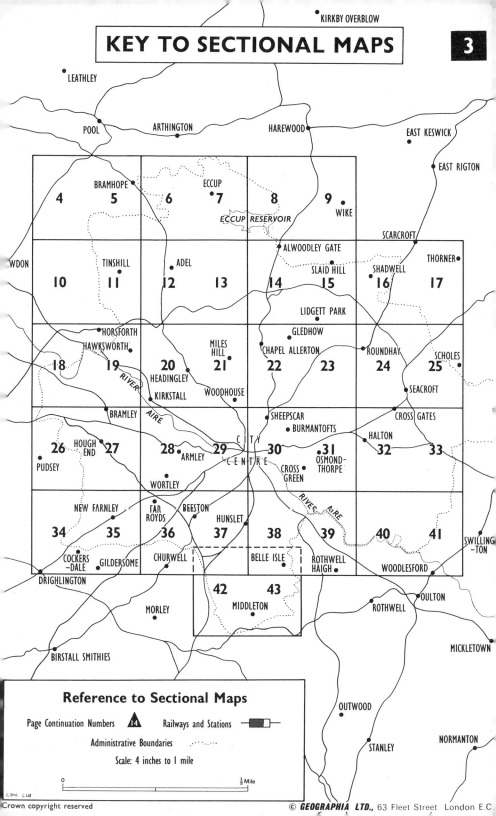

KEY TO SECTIONAL MAPS

3

KIRKBY OVERBLOW

LEATHLEY

POOL   ARTHINGTON   HAREWOOD   EAST KESWICK

EAST RIGTON

BRAMHOPE   ECCUP

4   5   6   7   8   9   WIKE

ECCUP RESERVOIR

SCARCROFT

WDON   TINSHILL   ADEL   ALWOODLEY GATE   SLAID HILL   SHADWELL   THORNER

10   11   12   13   14   15   16   17

LIDGETT PARK

HORSFORTH   MILES HILL   GLEDHOW   Roundhay   SCHOLES

HAWKSWORTH   CHAPEL ALLERTON

18   19   20   21   22   23   24   25

HEADINGLEY   SEACROFT

RIVER   KIRKSTALL   WOODHOUSE

BRAMLEY   AIRE   SHEEPSCAR   CROSS GATES

HOUGH   BURMANTOFTS   HALTON

26   END   27   28   29   CITY   30   31   32   33

PUDSEY   ARMLEY   CENTRE   OSMOND-   THORPE

CROSS GREEN

WORTLEY

NEW FARNLEY   FAR   BEESTON   RIVERS AIRE

ROYDS   HUNSLET

34   35   36   37   38   39   40   41

COCKERS   SWILLING-TON

-DALE   GILDERSOME   CHURWELL   BELLE ISLE   ROTHWELL   WOODLESFORD

DRIGHLINGTON   HAIGH

42   43

MORLEY   MIDDLETON   ROTHWELL   OULTON

BIRSTALL SMITHIES   MICKLETOWN

Reference to Sectional Maps

Page Continuation Numbers   **14**   Railways and Stations

Administrative Boundaries

Scale: 4 inches to 1 mile

0   ½ Mile

OUTWOOD

NORMANTON

STANLEY

Crown copyright reserved

© **GEOGRAPHIA LTD.,** 63 Fleet Street London E.C.

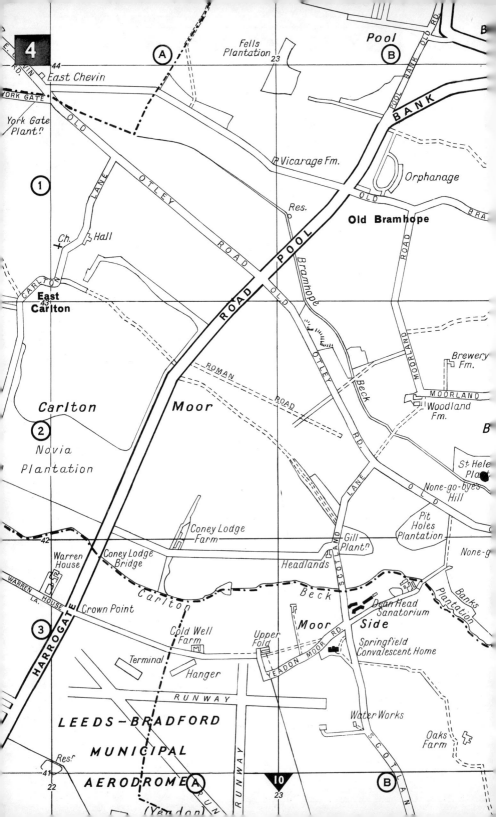

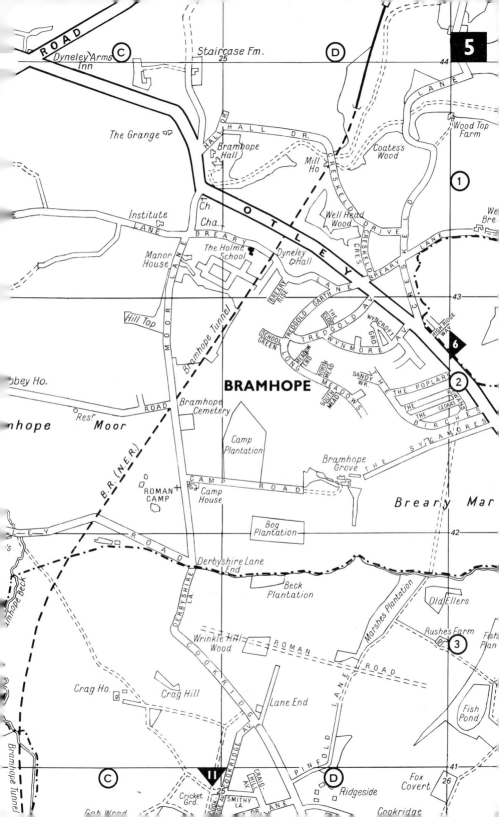

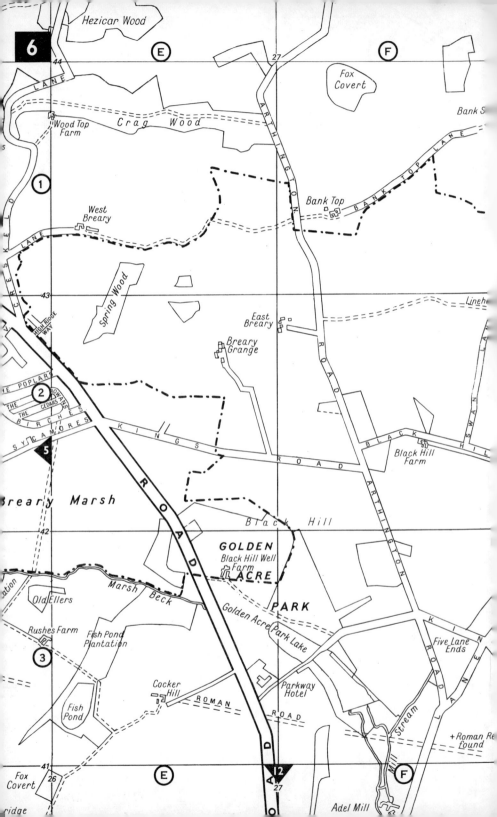

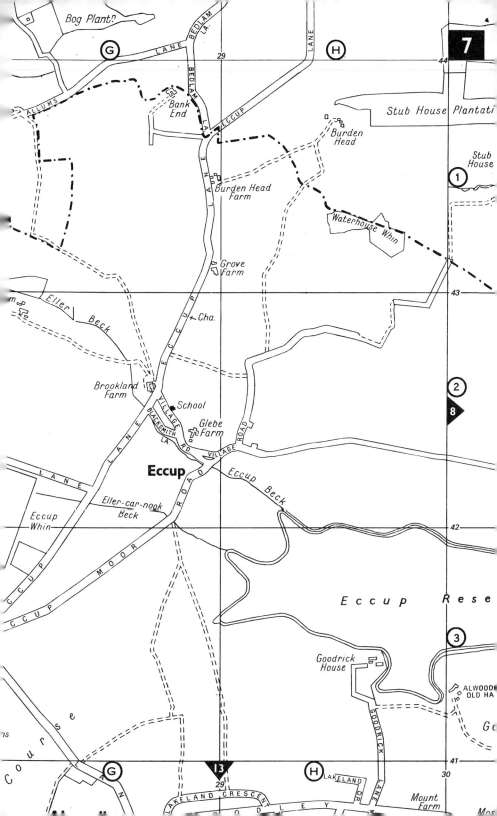

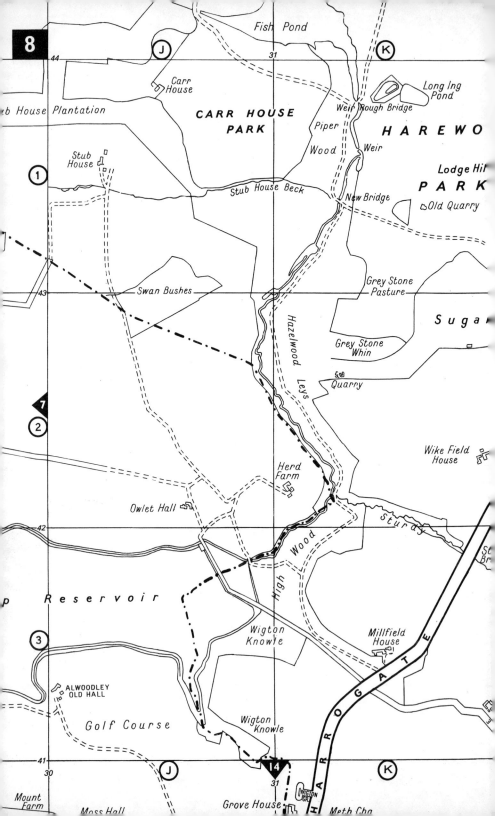

8

Fish Pond

Carr
House

Long Ing
Pond

**CARR HOUSE
PARK**

Weir Rough Bridge

Piper

H A R E W O

Wood

Weir

b House Plantation

Stub
House

Lodge Hill

P A R K

1

Old Quarry

Stub House Beck

New Bridge

Grey Stone
Pasture

Swan Bushes

43

S u g a

Hazelwood

Grey Stone
Whin

Leys

Quarry

7

Wike Field
House

2

Herd
Farm

Owlet Hall

Sturdy

42

High Wood

St
Br

Reservoir

Wigton
Knowle

Millfield
House

3

ALWOODLEY
OLD HALL

H A R R O G A T E

Golf Course

Wigton
Knowle

41
30

J

14

K

31

Grove House

WIGTON
GR

Meth Cha

Mount
Farm

Moss Hall

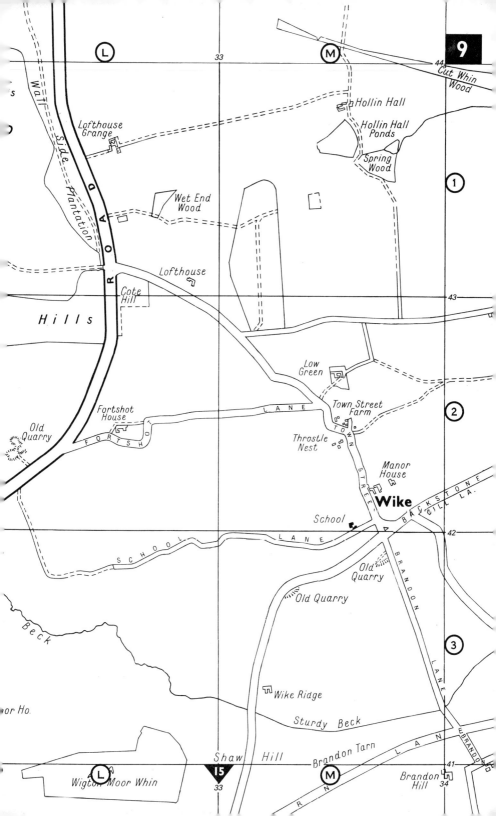

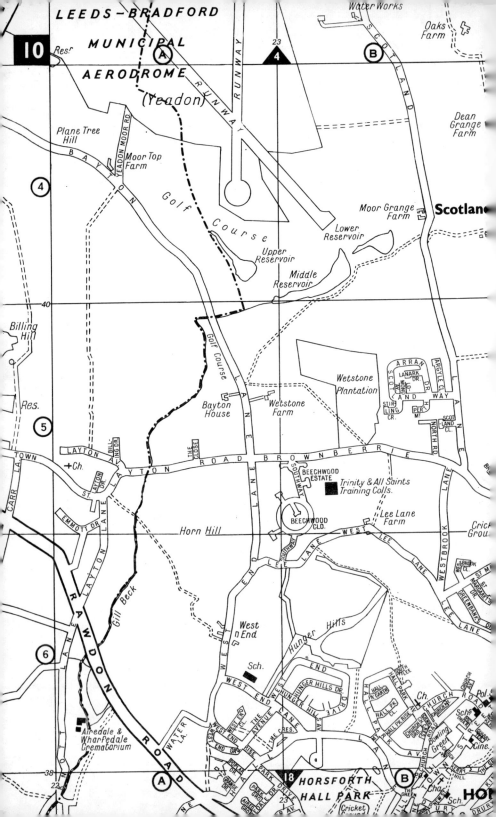

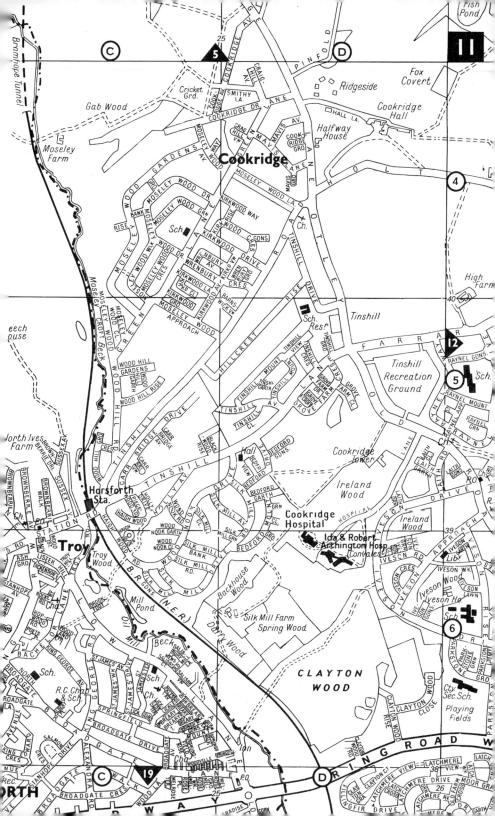

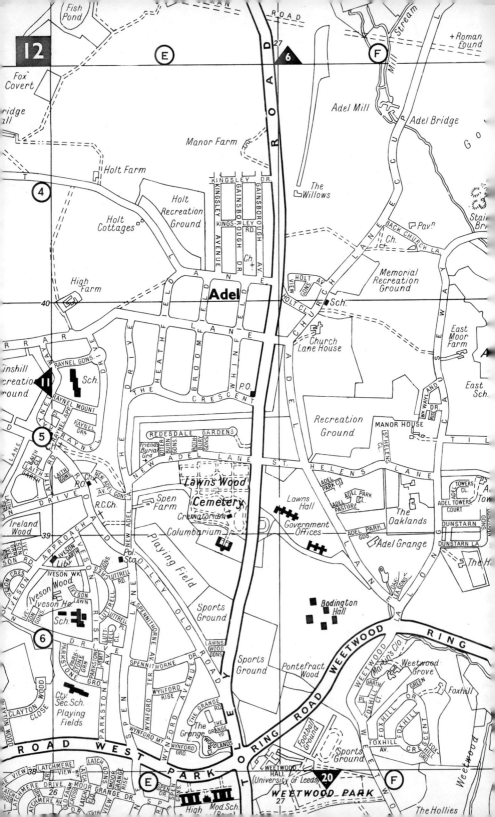

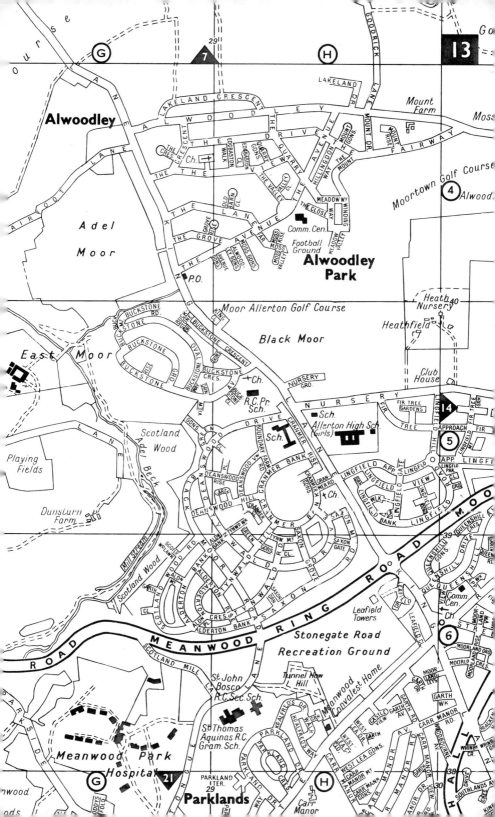

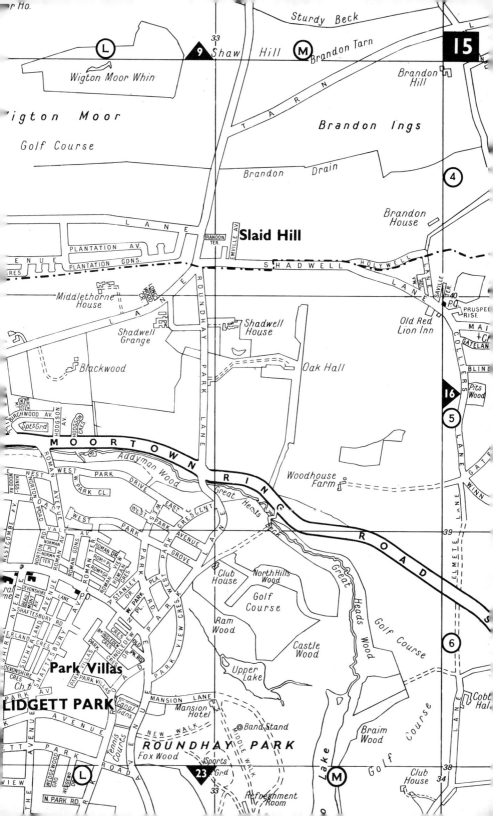

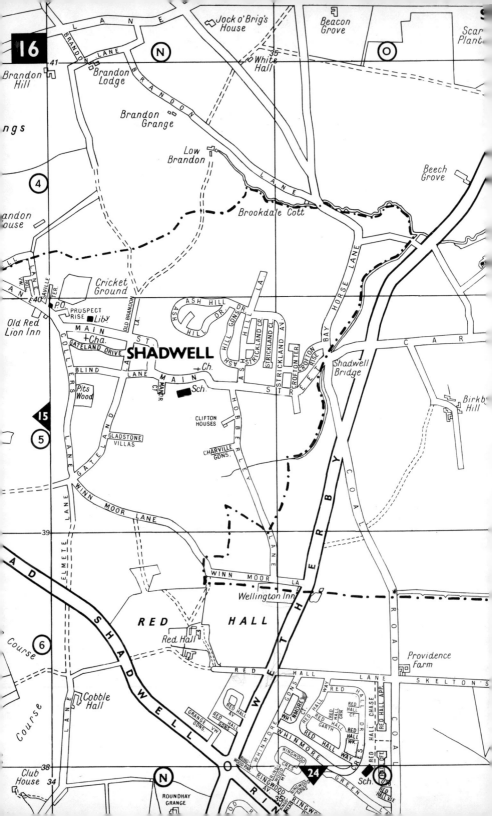

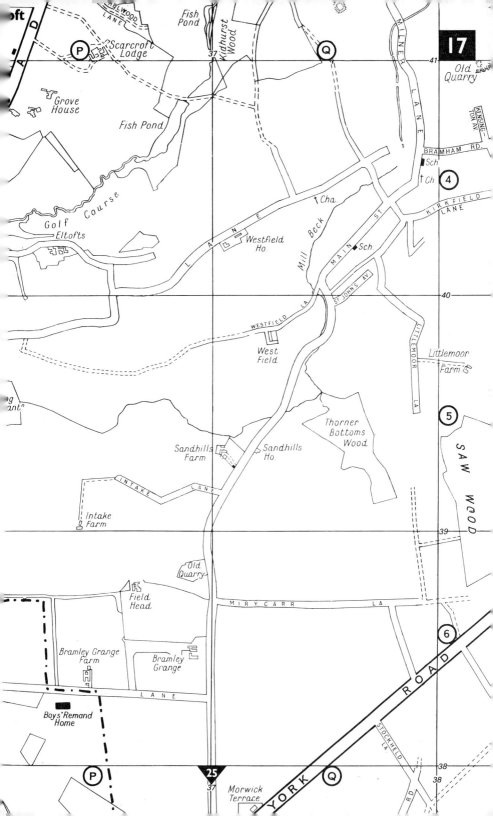

Old Quarry

P
Scarcroft Lodge
Fish Pond
Kidhurst Wood
37
41
Q

KENSINGTON AV.
BRAMHAM RD.
Sch
Ch
4
KIRKFIELD LANE

Grove House
Fish Pond

Cha.
Mill Beck
MAIN ST.
Sch

Golf Course
Eltofts
Westfield Ho.
L A N E

40

WESTFIELD LA.
West Field
St. JOHN'S AV.
Littlemoor Farm
LITTLEMOOR LA.

5

SAW WOOD

"ing ant"

Sandhills Farm
Sandhills Ho.
Thorner Bottoms Wood

INTAKE LANE

Intake Farm

39

Old Quarry
Field Head

MIRY CARR LA.
6

Bramley Grange Farm
Bramley Grange

L A N E

R O A D

Boys' Remand Home

STOCKHELD LA.

P
25
37
Q
38
38

Morwick Terrace
YORK ROAD

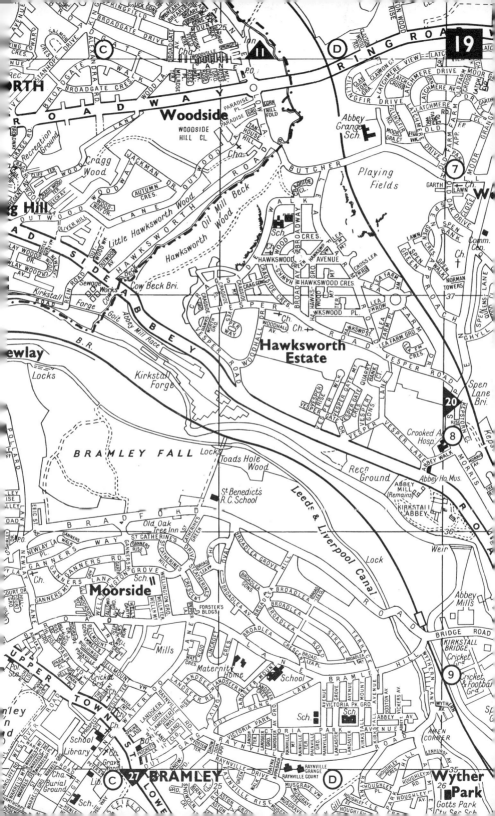

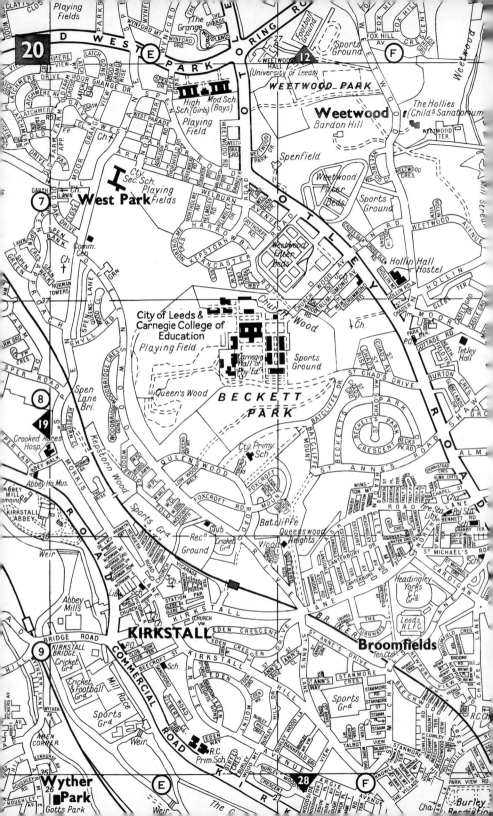

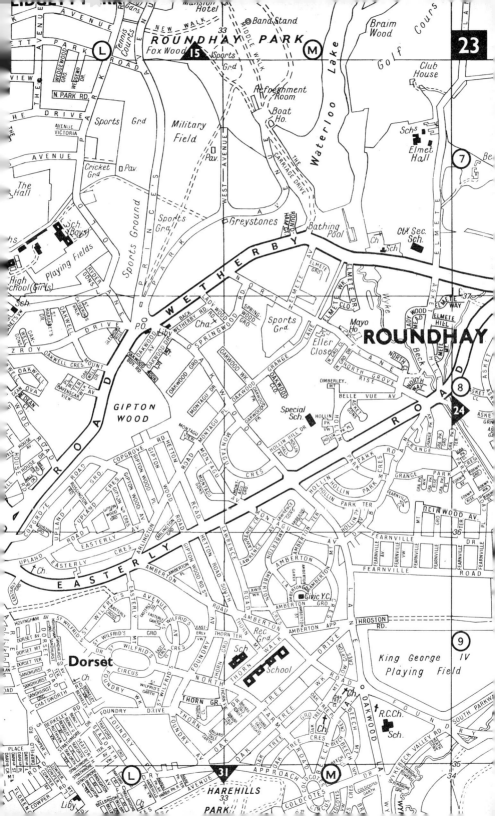

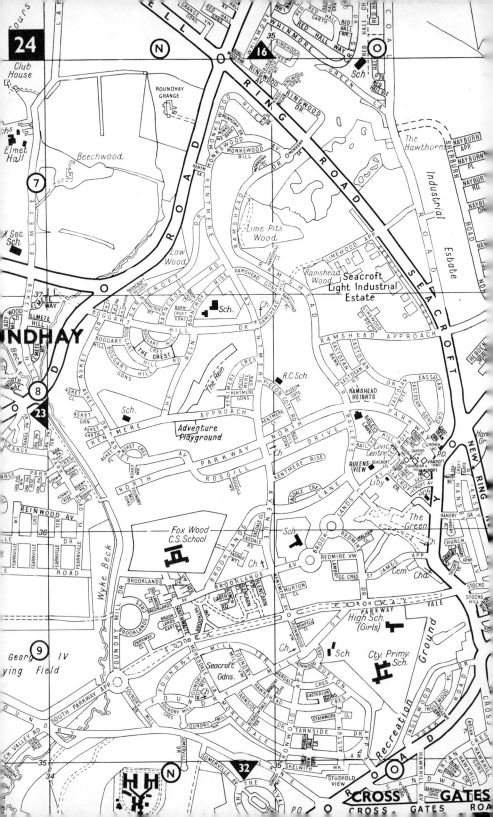

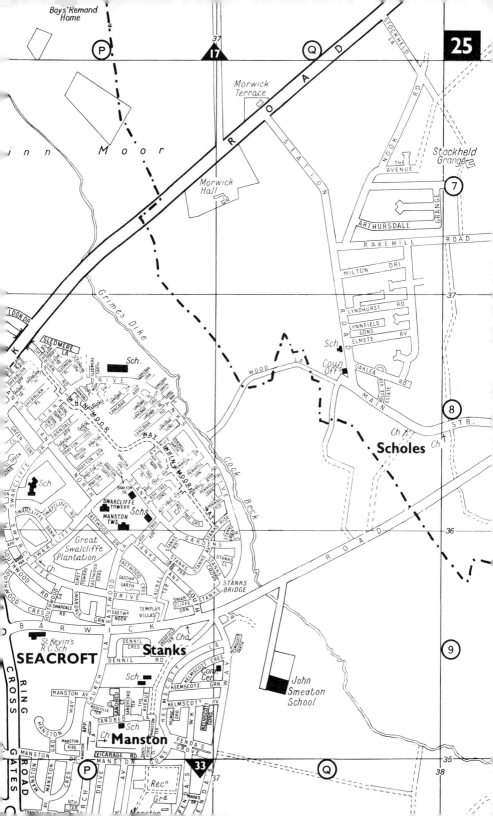

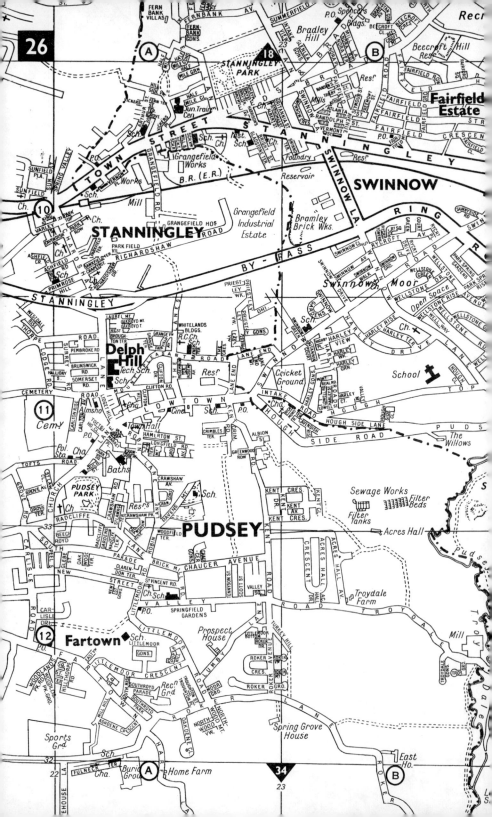

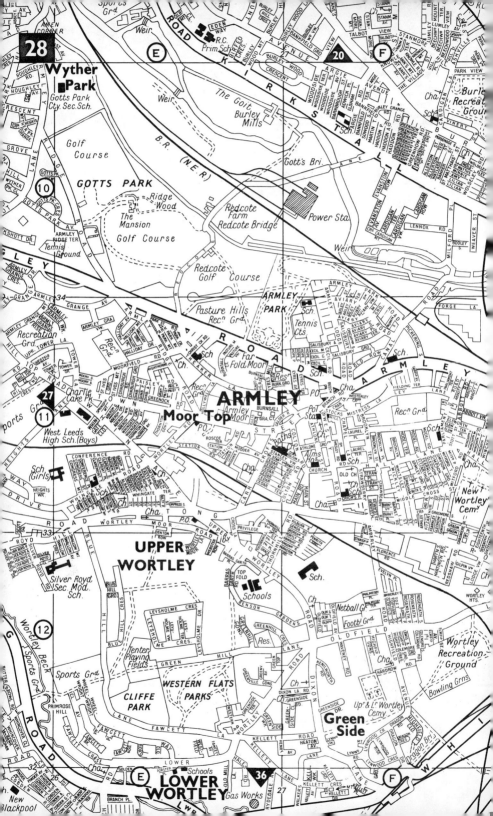

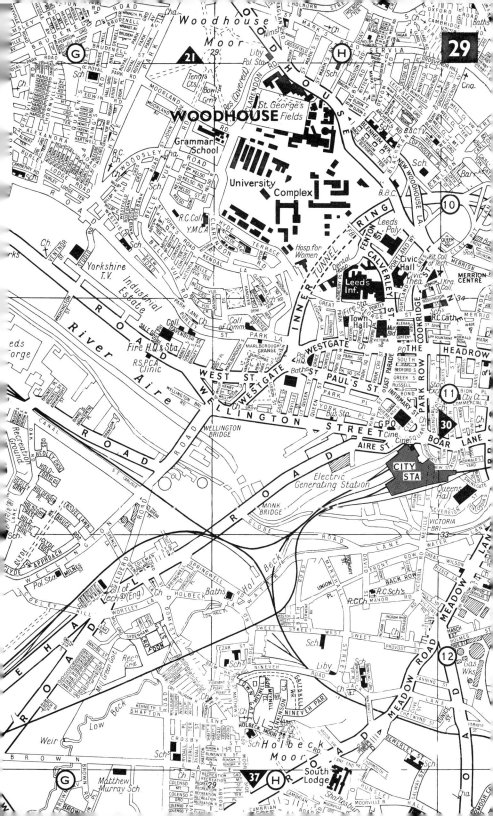

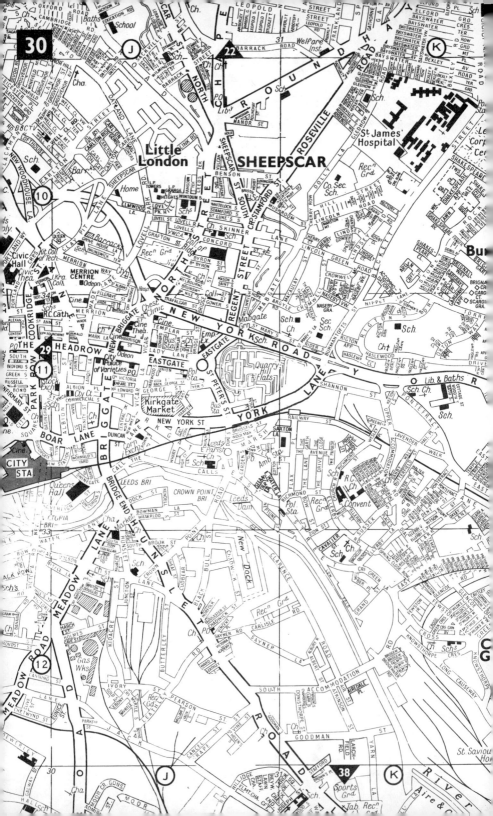

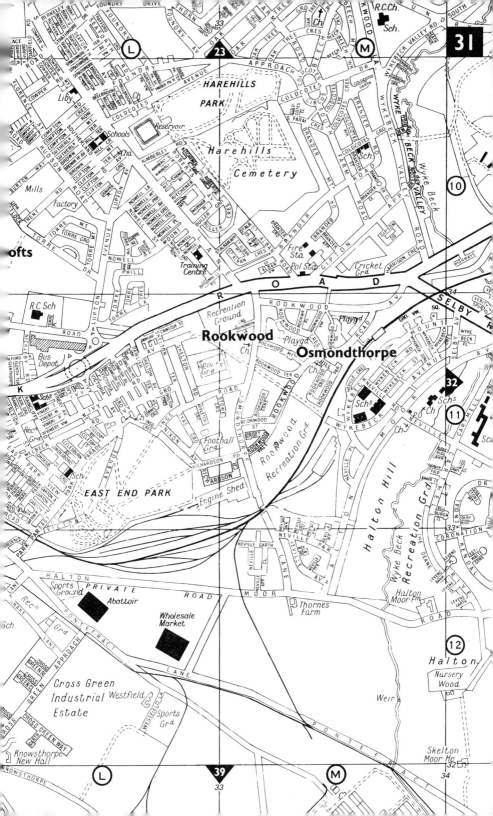

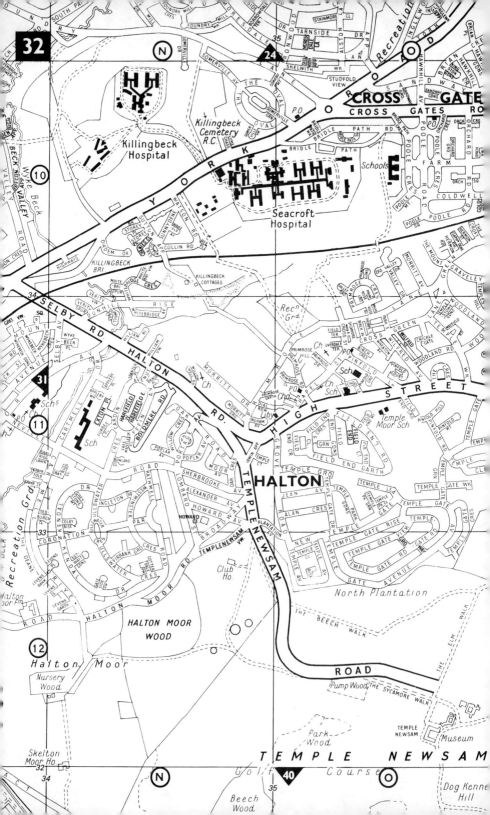

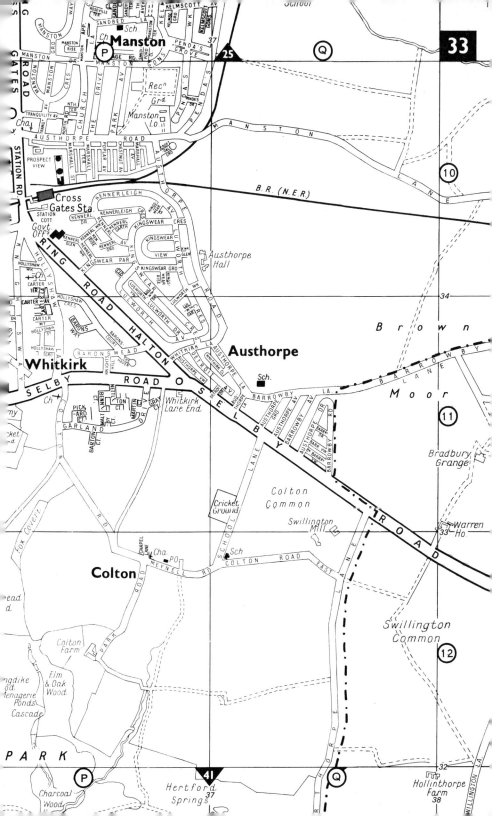

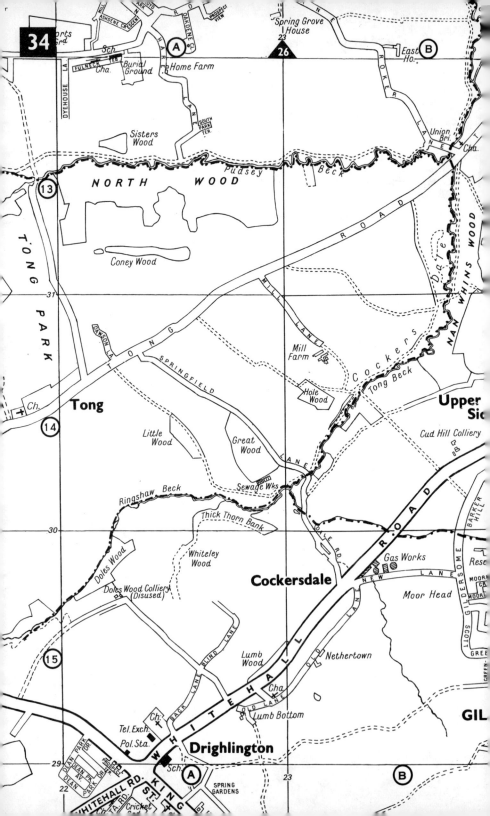

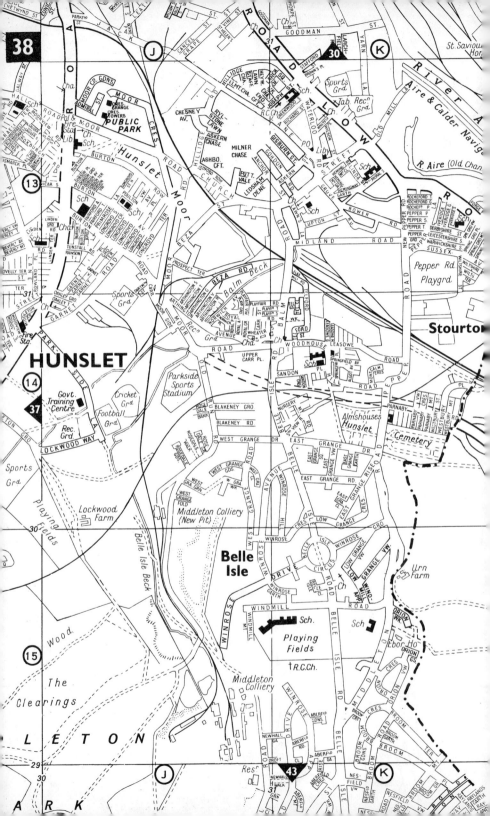

N
32 TEMPLE O NEWSAM

Park Wood

TEMPLE NEWSAM
Museum

Golf Course

Beech Wood

Dog Kenn Hill

Bell Wood

Spring Wood

The Old Walk

Dunstan Hills

13

Waterloo Main Colliery (Park Pits)

LANE

Wyke Beck Diversion

PONTEFRACT

Temple Thorpe Farm

31

Main Effluent Channel

Temple House Farm

Stapleton Ho. Farm

Thorpe Hall Farm

Beds

THORPE HALL (Remains)

39

14 Thorpe Hall Pasture

30 B.R. (L.M.R.)

Bullough Bri

Fishpond Lock

The Goit

Rothwell Haigh Collieries (Fanny Pit)

BULLOUGH LANE

15

Haigh Cottage

Haigh Farm

PICKPOCKET

Recn Grd

WOOD

TEMPLE FIRST AV.
Res

SECOND AV.
CRESCENT AV.
THIRD AV.
CROSS AV.

FOURTH AV.

FIFTH AV.
SIXTH AV.
SEVENTH AV.
EIGHTH AV.
NINTH AV.

HOLMESLEY AV.

HOLMESLEY GARTH

SPIBEY CR.
HIGH RIDGE PK.
HIGH RIDGE AV.

WILLANS LA.

HAIGH RIDGE

R O A D

John O'Gaunt's Industrial Estate

KINGS CHASE
KINGS MEAD

29
34 SOUTH VW.

DALMA TER.
WOODLANO LA.
PARK ST.
HAIGH RIDGE

Sch

N The Haigh Hosp

HOLMESLEY LANE

GIPSY LANE

Mill

RICHMOND
RICHMOND

KINGS CHASE
BR.

Cem

SPRING

EASTFIELD CRES.
EASTFIELD

LANE

35

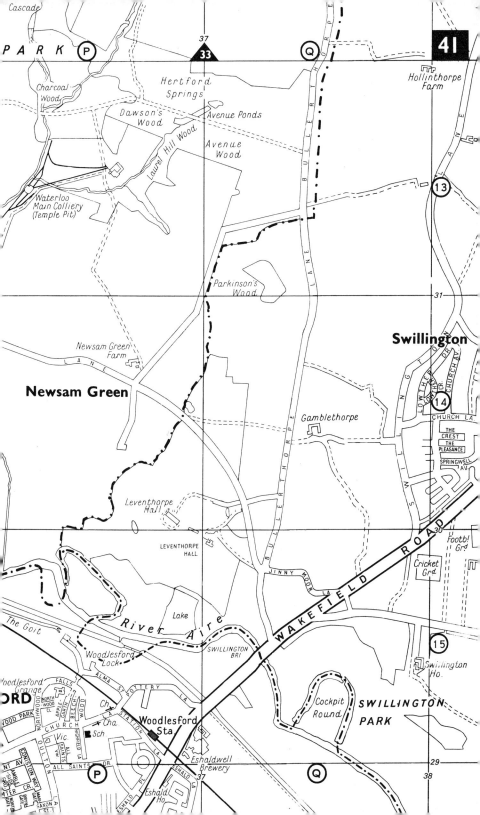

Cascade

P A R K

Ⓟ

37
▲ **33**

Ⓠ

🏠 *Hollinthorpe Farm*

Charcoal Wood

Hertford Springs

Dawson's Wood

Avenue Ponds

Laurel Hill Wood

Avenue Wood

B U L L E R T H O R P E   L A N E

⑬

Waterloo Main Colliery (Temple Pit)

Parkinson's Wood

31

Swillington

Newsam Green Farm

L A N E

B U L L E R T H O R P E   L A N E

Gamblethorpe

LOWTHER DR
CR
CHURCH AV

⑭

Newsam Green

CHURCH LA.

THE CREST
THE PLEASANCE

SPRINGWELL AV.

Leventhorpe Hall

LEVENTHORPE
HALL

JINNY MOOR   LA

30

Footbl Grd

Cricket Grd

W A K E F I E L D   R O A D

The Goit

River   Aire

Lake

⑮

Woodlesford Lock

SWILLINGTON BRI

🏠 *Swillington Ho.*

O R D

Woodlesford Grange

ALMA ST

POTTERY LA.

STATION LA.

Cockpit Round

SWILLINGTON PARK

NORTH WOOD
APPLE GARTH
BEECH

Ch.

Cha.

ALL SAINTS

Woodlesford Sta.

Vic.
Sch

HIGHFIELD

CHURCH LA

DOLTON

Ⓟ

ESHALD LA.

Eshaldwell Brewery

29

Ⓠ

38

Eshald Ho.

37

COLSTON WAY
AVSTON

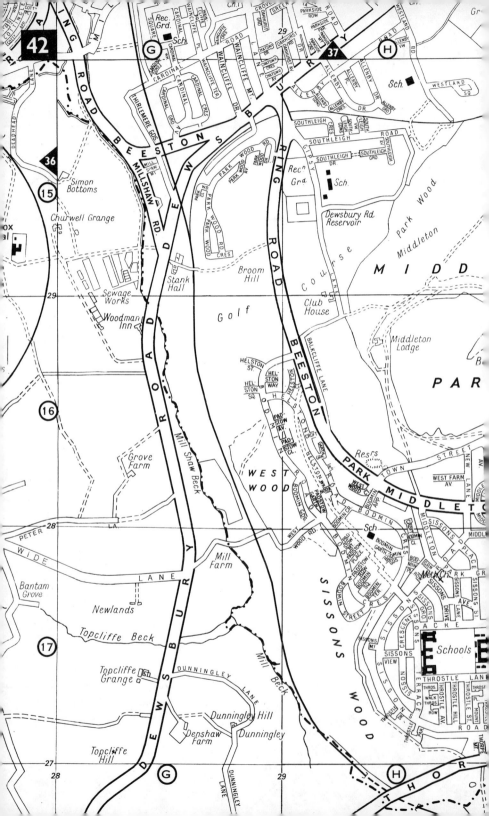

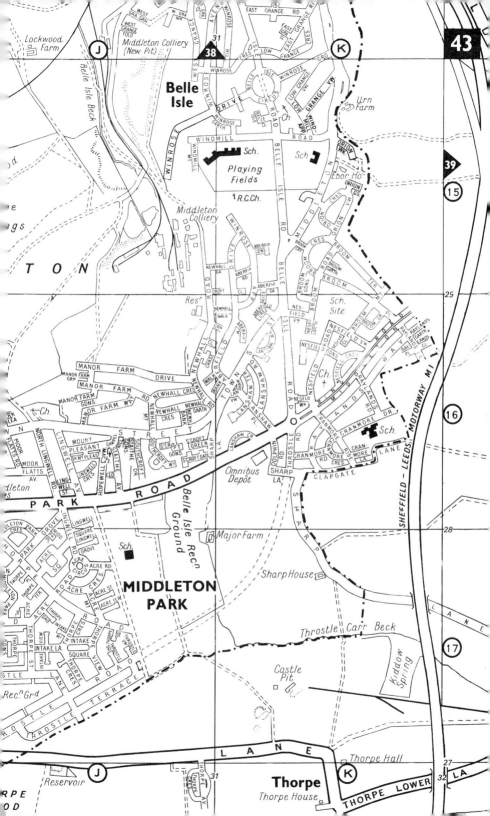

INDEX TO

GEOGRAPHIA

ATLAS OF

L E E D S

ABBREVIATIONS

App.—Approach.   Cl.—Close.   Gdns.—Gardens.   Mt.—Mount.   Ter.—Terrace.
Arc.—Arcade.   Cotts.—Cottages.   Gra.—Grange.   Par.—Parade.   Val.—Valley.
Av.—Avenue.   Ct.—Court.   Grn.—Green.   Pk.—Park.   Vw.—View.
Bldgs.—Buildings.   Cres.—Crescent.   Gro.—Grove.   Pl.—Place.   Wk.—Walk.
Bri.—Bridge.   Crs.—Cross.   Ho.—Houses.   Rd.—Road.   Wd.—Wood.
Circ.—Circus.   Dr.—Drive.   La.—Lane.   St.—Street.   Yd.—Yard.

NOTES

The figures and letters following a street name indicate the postal district and the square on the map where the name will be found. Thus Ashville Av. is in postal district 6 in map square E9, page 20.
A street name followed by the name of another street in italics does not appear on the Map but will be found adjoining or near the latter.

Abbey Av. 5	D9 19	
Abbey Mt. 5		
Raynville Rd.	D9 19	
Abbey Rd. 5	C8 19	
Abbey St. 3	G11 29	
Abbey Ter. 5	D9 19	
Abbey Wk. 5	D8 19	
Abbotsford Pl. 7	K9 22	
Abbott Pl. 12	F11 28	
Abbott Rd. 12	F11 28	
Abbott St. 12	F11 28	
Abbott Ter. 12	F11 28	
Abbott Vw. 12	F11 28	
Abercorn Pl. 11		
Chester Pl.	H13 37	
Abercorn St. 12		
Armley Rd.	F11 28	
Abercorn Ter. 12		
Armley Rd.	F11 28	
Aberdeen Dr. 12	E11 28	
Aberdeen Gro. 12	E11 28	
Aberdeen Rd. 12	E11 28	
Aberdeen Wk. 12	E11 28	
Aberfield Bank. 10	J16 43	
Aberfield Cl. 10	K15 38	
Aberfield Cres. 10	K16 43	
Aberfield Dr. 10	K16 43	
Aberfield Gdns. 10	L15 38	
Aberfield Garth. 10	K15 38	
Aberfield Gate. 10	K15 38	
Aberfield Mt. 10	K16 43	
Aberfield Rise. 10	K16 43	
Aberfield Rd. 10	K15 38	
Aberfield Wk. 10	J16 43	
Abyssinia Gro. 3		
St. John's Rd.	G10 29	
Abyssinia Rd. 3	G10 29	
Abyssinia St. 3	G10 29	
Abyssinia Ter. 3	G10 29	
Academy St. 10	J12 30	
Accommodation Rd.		
9	K10 30	
Acre Circus, 10	H17 42	
Acre Cres. 10	J17 43	
Acre Gro. 10	J17 43	
Acre Mt. 10	J17 43	
Acre Pl. 10	J17 43	
Acre Rd. 10 H17 to J17 43		
Acres Hall Av. 28	B12 26	
Acres Hall Cres. 28		
	B12 26	
Acres Hall Dr. 28	B12 26	
Acre Sq. 10	J17 43	
Acre St. 10	J17 43	
Acre Ter. 10	J17 43	
Acton St. 11		
Kiln St.	H13 37	
Ada Cres. 9	K11 30	
Ada Vw. 9	K11 30	
Adams St. 11		
Canning St.	H12 29	
Adams Wk. 6		
Moorland Rd.	G10 29	
Ada's Pl. 28		
Arthur St.	A10 26	
Addingham St. 12	E11 28	
Adel Lane, 16	F5 12	
Adel Park Gdns. 16	F5 12	
Adelphi St. 3	G11 29	
Adel Park Cl. 16	F5 12	
Adel Park Ct. 16	F5 12	
Adel Pasture, 16	F5 12	
Adel Towers Cl. 16	F5 12	
Adel Towers Ct. 16	F5 12	
Admiral St. 11	J13 38	
Admiral St. 11		
Dewsbury Rd.	H15 37	
Adwick Pl. 4	F10 28	
Airdale Prospect, 13	B9 18	
Airedale Cliff, 13	C8 19	
Airedale Dr. 18	A7 18	
Airedale Gro. 18	A7 18	
Airdale Pl. 1		
West St.	G11 29	
Airedale Pl. 9	K12 30	
Airedale Ter. 18	B7 18	
Airedale Ter. 10		
Wakefield Rd.	L14 39	
Airedale Vw. 13	A8 18	
Aire St. 1	H11 29	
Airlie Av. 8	K9 22	
Airlie Pl. 8	K9 22	
Air St. 10	K12 30	
Alan Cres. 15	O11 32	
Alaska Pl. 7	K8 22	
Albany Gro. 12	E11 28	
Albany Pl. 12	E11 28	
Albany St. 12	E11 28	
Albany Ter. 12	E11 28	
Alberta Av. 7	K7 22	
Alberta Dr. 7		
Montreal Av.	J8 22	
Albert Gro. 6	F8 20	
Albert Pl. 18	C6 11	
Albert St. 6	H9 21	
Albert St. 10	K12 30	
Albert St. 28	A10 26	
Albert Ter. 6	G10 29	
Albion Pl. 1	J11 30	
Albion St. 1	J11 30	
Albion St. 28	A11 26	

44

Albion Wk. 1		
Albion St.	J11	30
Albury Rd. 10	K12	30
Alcester Pl. 8		
Hill Top Av.	K9	22
Alcester Rd. 8	K9	22
Alcester Ter. 8		
Hill Top Av.	K9	22
Alder Hill Av. 7	H7	21
Alder Hill Gro. 7	H7	21
Alderon St. 12		
Lord St.	G12	29
Alderton Bank, 17	G6	13
Alderton Cres. 17	G6	13
Alderton Mt. 17	G6	13
Alderton Pl. 17	G6	13
Alderton Rise, 17	G6	13
Alexander Av. 15	N11	32
Alexander St. 1	H11	29
Alexandra Cres. 6	G10	29
Alexandra Gro. 6	G10	29
Alexandra Rd. 6	G10	29
Alexandra Rd. 18	C6	11
Alfred Pl. 7		
Camp Rd.	J10	30
Alfred St. 1	J11	30
Algot Av. 10	K14	38
Algot Gro. 10		
Springfield Av.	K14	38
Algot St. 10		
Springfield Av.	K14	38
Allenby Cres. 11	H15	37
Allenby Dr. 11	H15	37
Allenby Gdns. 11	H15	37
Allenby Gro. 11	H15	37
Allenby Pl. 11	H15	37
Allenby Rd. 11	H15	37
Allenby Vw. 11	H14	37
Allerton Av. 17	J6	14
Allerton Grange, 17	J6	14
Allerton Grange Av. 17	K6	14
Allerton Grange Cres. 17	J7	22
Allerton Grange Dr. 17	J7	22
Allerton Grange Gdns. 17	J7	22
Allerton Grange Rise, 17	J7	22
Allerton Grange Vale, 17	J7	22
Allerton Grange Way, 17	J7	22
Allerton Gro. 17	J6	14
Allerton Hill, 7	J7	22
Allerton Pk. 7	J7	22
Allerton Pl. 17		
Allerton Gro.	J6	14
Allerton St. 4	F10	28
Allerton Ter. 4	F10	28
All Saints Dr. 26	P16	41
All Saints Vw. 26	P15	41
Allums Lane, 16	G1	7
Alma Cotts. 6	F8	20
Alma Gro. 9	K10	30
Alma Rd. 6	F8	20
Alma St. 26	P15	41
Alma Ter. 26	M16	39
Aloe St. 10		
Orchard St.	K12	30
Alpha Gro. 11		
Alpha St.	J12	30
Alpha Ter. 11		
Alpha St.	J12	30
Alston La. 14	O9	24
Altofts Pl. 11	H13	37
Altofts St. 11		
Cambrian Rd.	H13	37
Altofts Ter. 11	H13	37
Alwoodley Gdns. 17	H4	13
Alwoodley Lane, 17	G4	13
Amberley Gro. 7	J9	22
Amberley Rd. 12	F11	28
Amberley St. 12	F12	28
Amberley Ter. 7	J9	22
Amberton App. 8	M9	23
Amberton Cl. 8	M8	23

Amberton Cres. 8	M9	23
Amberton Gdns. 8	M9	23
Amberton Garth, 8	M9	23
Amberton Gro. 8	M9	23
Amberton La. 8	M9	23
Amberton Mt. 8	M9	23
Amberton Pl. 8	L9	23
Amberton Rd. 8	L9	23
Amberton St. 8	M9	23
Amberton Ter. 8	M9	23
Amen Corner, 12	D9	19
Amy St. 4	F10	28
Ancaster Cres, 16	E7	20
Ancaster Rd. 16	E7	20
Ancaster Vw. 16	E7	20
Anchor St. 10	J13	38
Anderson Av. 9		
Dolly La.	K10	30
Anderson Mt. 8	K10	30
Andrew St. 10		
Holdsworth St.	J12	30
Angel Inn Yd. 1		
Briggate	J11	30
Appleby Pl. 15	N11	32
Appleby Wk. 15	N11	32
Applegarth, 26	P15	41
Appleton Cl. 9	K11	30
Appleton Ct. 9	K11	30
Appleton Garth, 9	K11	30
Appleton Gro. 9	L11	31
Appleton Sq. 9	K11	30
Appleton Way, 9	K11	30
Appleyard La. 12		
Town St.	E11	28
Archery Pl. 2	H10	29
Archery Rd. 2	H10	29
Archery St. 2		
Archery Rd.	H10	29
Archery Ter. 2		
Archery Rd.	H10	29
Archie Pl. 9	K12	30
Arch Rd. 12	G12	29
Arch Ter. 12	G12	29
Arden Croft, 9	K10	30
Arden Pl. 9	K10	30
Argie Av. 4	E9	20
Argie Gdns. 4		
Burley Rd.	F10	28
Argie Rd. 4	F10	28
Argie Ter. 4	F10	28
Argyle Cl. 18	B5	10
Argyle Rd. 9	J11	30
Arksey Ter. 12		
Aviary Rd.	F11	28
Arkwright St. 12	G11	29
Arley Gro. 12		
Aviary Rd.	F11	28
Arley Pl. 12	F11	28
Arley St. 12	F11	28
Arley Ter. 12	F11	28
Arlington Gro. 8	L8	23
Arlington Rd. 8	L9	23
Armenia Gro. 7		
Manor St.	J10	30
Armenia Ter. 7		
Manor St.	J10	30
Armley Grange Av. 12	D10	27
Armley Grange Cres. 12	D10	27
Armley Grange Dr. 12	D11	27
Armley Grange Mt. 12	D11	27
Armley Grange Oval, 12	D10	27
Armley Grange Rise, 12	D11	27
Armley Grange Vw. 12	E11	28
Armley Grange Wk. 12	D11	27
Armley Grove Pl. 12	F11	28
Armley Lodge Rd. 12	F10	28
Armley Park Rd. 12	F10	28
Armley Ridge Rd. 12 D9 to E11		19

Armley Ridge Ter. 12	E10	28
Armley Rd. 12	F11	28
Arncliffe Rd. 16	E7	20
Arran Dr. 18	B5	10
Arthington Av. 10	J14	38
Arthington Gro. 10	J14	38
Arthington Rd. 10	J14	38
Arthington Rd. Bra.		
and 16	E1 to F3	6
Arthington St. 10	J14	38
Arthington Ter. 10	J14	38
Arthington Vw. 10	J14	38
Arthursdale Grange, 14	Q7	25
Arthur St. 28	A10	26
Artillery Pl. 7		
Roundhay Rd.	K9	22
Artillery Ter. 7		
Roundhay Rd.	K9	22
Arundel Gro. 8	K10	30
Arundel Mt. 8	K10	30
Arundel Pl. 8	K10	30
Arundel St. 8	K10	30
Arundel Ter. 8	K10	30
Ascot Av. 9	K11	30
Ascot Pl. 9		
Ascot St.	K11	30
Ascot St. 9	K11	30
Ascot Ter. 9	K11	30
Ashbourne Croft, 10	J13	28
Ashby Cres. 13	C10	27
Ashby Ter. 13	C10	27
Ashby Vw. 13	C10	27
Ash Cres, 6	F8	20
Ashdene, 12	C13	35
Ashdene Cres, 28	A12	26
Ashfield, 12	E13	36
Ashfield Cres. 28	A10	26
Ashfield Gro. 28	A10	26
Ashfield Rd. 28	A10	26
Ashfield Ter. 15	P9	25
Ashford St. 11		
Bewerly St.	H12	29
Ash Gdns. 6	F8	20
Ash Gro. 6	G9	21
Ash Gro. 18	C6	11
Ash Hill Dr. 17	N5	16
Ash Hill Gdns. 17	N5	16
Ash Hill La. 17	N5	16
Ashlea Gro. 13		
Ashlea Rd.	C9	19
Ashlea Pl. 13		
Ashlea Rd.	C9	19
Ashlea Rd. 13	C9	19
Ashlea St. 13		
Ashlea Rd.	C9	19
Ashleigh Rd. 16	E7	20
Ashley Av. 9		
Ashley Rd.	K10	30
Ashley Rd. 9	K10	30
Ashley Rd. 12		
Wortley Rd.	E12	28
Ashley Ter. 9		
Ashley Rd.	K10	30
Ash Rd. 6 F9 to F8		20
Ash Ter. 6	F8	20
Ashton Av. 8	K9	22
Ashton Gro. 8	K10	30
Ashton Mt. 8	K10	30
Ashton Pl. 8	K10	30
Ashton Rd. 8	L9	23
Aston St. 8	K9	22
Aston Ter. 8	K10	30
Ashton Vw. 8	K10	30
Ash Tree App. 14	P9	25
Ash Tree Bank, 14	P8	25
Ash Tree Cl. 14	P8	25
Ash Tree Gdns. 14	P8	25
Ash Tree Gro. 14	P8	25
Ash Tree Vw. 14	P8	25
Ash Tree Wk. 14	P8	25
Ash Vw. 6	F8	20
Ashville Av. 6	F9	20
Ashville Gro. 6	F9	20
Ashville Rd. 4 and 6	F9	20
Ashville Ter. 6	F9	20
Ashville Vw. 6		
Cardigan Rd.	F9	20

Baptist St. 12		
Wellington Rd.	G12	29
Barclay St. 7		
North St.	J10	30
Barden Gro. 12		
Whingate Rd.	E11	28
Barden Mt. 12	E11	28
Barden Pl. 12	E11	28
Barden St. 12	E11	28
Barden Ter. 12	E11	28
Barfield Cres. 17	K4	14
Barker Hill, 27	B14	34
Barker Pl. 13		
Scarborough		
Junction	C10	27
Barkers Pl. 13	C10	27
Barkly Av. 11	H14	37
Barkly Dr. 11	H14	37
Barkly Gro. 11	H14	37
Barkly Par. 11	H14	37
Barkly Rd. 11	G14	37
Barkly St. 11	H14	37
Barkly Ter. 11	H14	37
Barkston Pl. 11	G12	29
Barkston Ter. 11	G12	29
Bar Lane, 18	A7	18
Barleycorn St. 11		
Whitehall Rd.	G12	29
Barmouth Rd. 10		
Balm Rd.	K14	38
Barnbrough St. 4	F10	28
Barncroft Cl. 14	N7	24
Barncroft Dr. 14		
	N8 to N7	24
Barncroft Gdns. 14	N8	24
Barncroft Heights, 14		
	N7	24
Barncroft Mt. 14	N8	24
Barncroft Rise, 14		
	N8 to N7	24
Barncroft Rd. 14		
	N8 to N7	24
Barnet Mt. 12		
Strawberry Rd.	F11	28
Barnet Pl. 12	F11	28
Barnet Rd. 12	F11	28
Barnet Vw. 12	F11	28
Barnswick Vw. 16 ·	D4	11
Barnthorpe Cres. 17	J7	22
Baron Cl. 11	H13	37
Baronscourt, 15	P11	33
Baronsmead, 15	P11	33
Barons Way, 15	P11	33
Barrack Rd. 7	J9	22
Barrack St. 7	J10	30
Barras Garth Pl. 12	E12	28
Barras Garth Rd. 12		
	E12	28
Barras Pl. 12		
Upper Wortley Rd.	E11	28
Barras St. 12		
Upper Wortley Rd.	E11	28
Barras Ter. 12		
Upper Wortley Rd.	E11	28
Barratt St. 10		
Holdsworth St.	J12	30
Barrister St. 9		
Beckett St.	K10	30
Barrowby Av. 15	Q11	33
Barrowby Dr. 15	Q11	33
Barrowby La. 15		
	Q11 to S11	33
Barrowby Rd. 15	Q11	33
Barrowby Vw. 15	Q11	33
Barstow St. 11	J12	30
Barthorpe Av. 17	H7	21
Barton Ct. 15	P11	33
Barton Gro. 11	H13	37
Barton Mt. 11		
Barton Rd.	H13	37
Barton Pl. 11		
Barton Rd.	H13	37
Barton Rd. 11	H13	37
Barton Ter. 11	H13	37
Barton Vw. 11		
Barton Rd.	H13	37
Barwick Rd. 15		
	P9 to S8	25

Basinghall Sq. 1		
Basinghall St.	·H11	29
Basinghall St. 1	H11	29
Batcliffe Dr. 6	F8	20
Batcliffe Mt. 6	F8	20
Bateson St. 10	L14	39
Bath Av. 13	C10	27
Bath Cl. 13	C10	27
Bath Gro. 13		
Bath Rd.	C10	27
Bath Lane, 13	C10	27
Bath Pl. 13	C10	27
Bath Rd. 11	H12	29
Bath Rd. 13	C10	27
Bath St. 9	K11	30
Bath St. 13	C10	27
Bath Vw. 13	C10	27
Battersea Pl. 10		
Thwaite La.	L13	39
Battersea St. 10		
Thwaite La.	L13	39
Bawn Av. 12	D12	27
Bawn App. 12	D12	27
Bawn Chase, 12	D12	27
Bawn Dr. 12	D12	27
Bawn Gdns. 12	D12	27
Bawn La. 12	D12	27
Bawn Pl. 12	D12	27
Bawn Vale, 12	D12	27
Bawn Wk. 12	D12	27
Bay Horse La. 17	O5	17
Bay Horse Yd. 1		
Briggate	J11	30
Bayswater Av. 8	K9	22
Bayswater Cres. 8	K9	22
Bayswater Gro. 8	K9	22
Bayswater Mt. 8	K9	22
Bayswater Pl. 8	K9	22
Bayswater Rd. 8	K9	22
Bayswater Row, 8	K9	22
Bayswater St. 8	K9	22
Bayswater Ter. 8	K9	22
Bayswater Vw. 8	K9	22
Bayton Lane, 19	A4	10
Beaconsfield Pl. 5		
Victoria Rd.	E9	20
Beamsley Gro. 6		
Harold Rd.	G10	29
Beamsley Mt. 6		
Royal Park Rd.	G10	29
Beamsley Pl. 6		
Harold Gro.	G10	29
Beamsley Ter. 6		
Royal Park Rd.	G10	29
Beatrice Ter. 10		
Woodhouse Hill Rd.	K14	38
Beaumont Av. 8	L6	15
Beckett Pk.		
Otley Rd.	D2	5
Beckett St. 9	K10	30
Beckett's Park Cres.		
6	F8	20
Beckett's Park Dr. 6	F8	20
Beckett's Park Rd. 6	F8	20
Beckhill App. 7	H8	21
Beckhill Av. 7	H8	21
Beckhill Chase, 7	H8	21
Beckhill Av.		
Beckhill Cl. 7	H8	21
Beckhill Dr. 7	H7	21
Beckhill Fold. 7	H7	21
Beckhill Lawn, 7	H8	21
Beckhill Pl. 7	H7	21
Beckhill Vale, 7	H7	21
Beckhill Way, 7	H8	21
Beck Rd. 8	K9	22
Bedford Cl. 16	D5	11
Bedford Dr. 16	D5	11
Bedford Gdns. 16	D5	11
Bedford Garth, 16	D5	11
Bedford Grn. 16	D5	11
Bedford Gro. 16	D6	11
Bedford Mt. 16	D6	11
Bedford Row, 10	J13	38
Bedford St. 1	H11	29
Bedford Ter. 7	J10	30
Bedford Vw. 16	D5	11
Bedlam La. 16	G1	7

Beech Av. 12		
Stanningley Rd.	D10	27
Beech Av. 18	C7	19
Beech Cl. 9	M9	23
Beech Cres. 9	M9	23
Beech Dr. 12	F11	28
Beechfield, 12	C13	35
Beech Gro. 10		
Pontefract Rd.	K13	29
Beech Gro. 12		
Stanningley Rd.	D10	27
Beech La. 9	M9	23
Beech Mt. 9	M9	23
Beech Royd, 28	A12	26
Beech Ter. 12		
Stanningley Rd.	D10	27
Beech Vw. 12		
Stanningley Rd.	D10	27
Beech Wk. 9	M9	23
Beech Wk. The, 15	O12	33
Beechwood, 26	P15	41
Beechwood Av. 4		
Beechwood Cres.	F9	20
Beechwood Cl. 18	B5	10
Beechwood Cres. 4	F9	20
Beechwoood Estate,		
18	B5	10
Beechwood Gro. 4		
Beechwood Cres.	F9	20
Beechwood Mt. 4	F9	20
Beechwood Pl. 4		
Lumley Av.	F9	20
Beechwood Rd. 4		
Beechwood Cres.	F9	20
Beechwood Row, 4		
Beechwood Cres.	F9	20
Beechwood St. 4		
Beechwood Cres.	F9	20
Beechwood Ter. 4	F9	20
Beechwood Vw. 4	F9	20
Beechwood Wk. 4		
Beechwood Cres.	F9	20
Beecroft Cl. 13	B9	18
Beecroft Cres. 13	B9	18
Beecroft Gro. 7		
Leopold St.	J9	22
Beecroft Mt. 13	B9	18
Beecroft St. 5	F9	20
Beeston Pk. 10	H16	42
Beeston Park Croft,		
11	G14	37
Beeston Park Garth,		
11	G14	37
Beeston Park Gro.		
11	G14	37
Beeston Park Pl 11	G14	37
Beeston Park Ter. 11	G14	37
Beeston Rd. 11	G14	37
Beeston Rd. 11	H12	29
Beeston Royds, 11		
Gelderd Rd.	H12	29
Beevers Ter. 11		
Coupland St.	H13	37
Belgrave St. 2	J10	30
Belgrave Ter. 2		
·Belgrave St.	J10	30
Belinda St. 10	K13	38
Bellbrooke Av. 9	L10	31
Bellbrooke Gro. 9	L10	31
Bellbrooke Pl. 9	L10	31
Bellbrooke St. 9	L10	31
Belle Isle Circ. 10	K15	38
Belle Isle Cl. 10	K15	38
Belle Isle Par. 10	K14	38
Belle Isle Rd. 10	K14	38
Belle Vue Av. 3		
Belle Vue Rd.	G10	29
Belle Vue Av. 8	M8	23
Belle Vue Cres. 3	G10	29
Belle Vue Estate,		
14	Q8	25
Belle Vue Gro. 3	G10	29
Belle Vue Lawn, 3		
Belle Vue Rd.	G10	29
Belle Vue Pl. 3		
Belle Vue Cres.	G10	29
Belle Vue Pl. 11		
South Ridge St.	H13	37

Belle Vue Rd. 3 and 6 G10 29
Belle Vue St. 3
Belle Vue Rd. G10 29
Belle Vue Ter. 3 G10 29
Bell Lane, 7 J9 22
Bell Lane, 13 C9 19
Bell Mt. 13 *Bell La.* C9 19
Bellmount Gdns. 13 C9 19
Bellmount Gro. 13 C9 19
Bellmount Vw. 13 C9 19
Bell Rd. 13 C9 19
Belmont Cres. 10
Spring Grove St. K13 38
Belmont Gro. 2
Clarendon Rd. H10 29
Belmont Mt. 10
Spring Grove St. K13 38
Belmont Vw. 10
Spring Grove St. K13 38
Belvedere Av. 11
Harlech Rd. H14 37
Belvedere Av. 17 J5 14
Belvedere Gdns. 17
Belvedere Rd. K5 14
Belvedere Gro. 13
Bell La. C9 19
Belvedere Gro. 17 J5 14
Belvedere Rd. 17 J5 14
Belvedere Ter. 11
Harlech Rd. H14 37
Belvedere Vw. 17 K5 14
Bennett Ct. 15 P11 33
Bennett Rd. 6 F8 20
Benson Gdns. 12 E12 28
Benson Pl. 10
Woodhouse Hill Rd. K14 38
Benson St. 7 J10 30
Bentcliffe Av. 17 J6 14
Bentcliffe Cl. 17 K6 14
Bentcliffe Dr. 17 K6 14
Bentcliffe Gdns. 17 K6 14
Bentcliffe Gro. 17 K6 14
Bentcliffe La. 17 J6 14
Bentcliffe Mt. 17 K6 14
Bentley Gro. 6 G8 21
Bentley Lane, 6 and 7 G8 21
Bentley Mt. 6 G8 21
Bentley Par. 6 G8 21
Berkeley Av. 8 L9 23
Berkeley Cres. 8
Berkeley St. L9 23
Berkeley Gro. 8 L9 23
Berkeley Mt. 8
Strathmore Dr. L9 23
Berkeley Rd. 8 L9 23
Berkeley St. 8 L9 23
Berkeley Ter. 8 L9 23
Berkeley Vw. 8 L9 23
Berking Av. 9 K11 30
Berking Pl. 9 K11 30
Berking Row, 9
Berking Av. K11 30
Berking St. 9
Berking Av. K11 30
Berking Ter. 9
Temple View Rd. K11 30
Bernard St. 2
Union St. J11 30
Bertha Cres. 9 K11 30
Bertha Gro. 9 K11 30
Bertha Mt. 9 K11 30
Bertha Pl. 9 K11 30
Bertha St. 9 K11 30
Bertha Vw. 9 K11 30
Bessbrook Mt. 10
Church St. J13 38
Bessbrook Pl. 10
Church St. J13 38
Bessbrook St. 10
Church St. J13 38
Bessbrook Ter. 10
Church St. J13 38
Beulah Gro. 6
Beulah Ter. H9 21
Beulah Mt. 6
Christopher Rd. H9 21

Beulah St. 6
Beulah Ter. H9 21
Beulah Ter. 27 C15 35
Beulah Ter. 6 H9 21
Beulah Ter. 15
Austhorpe Rd. P10 33
Beulah Vw. 6
Christopher Rd. H9 21
Beverley Av. 11 H13 37
Beverley Mt. 11 H13 37
Beverley Ter. 11 H13 37
Beverley Vw. 11 H13 37
Bewerley St. 11 H12 29
Bexley Av. 8 K10 30
Bexley Gro. 8 K10 30
Bexley Mt. 8 K10 30
Bexley Pl. 8 K10 30
Bexley Rd. 8 K10 30
Bexley Ter. 8 K10 30
Bexley Vw. 8
Bexley Gro. K10 30
Beza Rd. 10 J13 38
Beza St. 10 J13 38
Bideford Av. 8 K6 14
Billey Lane, 12 D12 27
Billingbauk Dr. 13 C10 27
Billingbauk Ter. 13
Brighton Gro. C10 27
Billing Dr. 19 A5 10
Bilton Ter. 11
Galway St. H13 37
Bingley St. 3 G11 29
Binks St. 12 G11 29
Birch Av. 15 O11 33
Birch Cres. 15 O11 33
Birches, The, 16 D2 5
Birchfield Av. 27 C16 35
Birchfield Pl. 27 C16 35
Birchwood Av. 17 L5 15
Birchwood Hill, 17 L5 15
Birchwood Mt. 17 L5 15
Birfed Cres. 4 E9 20
Bischoff's Yd. 2
North St. J10 30
Bishopsgate St. 1 H11 29
Black Bull St. 10 J12 30
Black Hill Lane, 16 F2 6
Blackman Lane,
2 and 7 H11 29
Black Moor Rd. 17 G5 13
Blackpool Gro. 12
Cow Close Rd. D12 27
Blackpool St. 12 E13 36
Blackpool Ter. 12 E13 36
Blacksmith Lane, 16 G2 7
Blackwood Rise, 16 C15 35
Blairsville Gdns. 13 B9 18
Blairsville Gro. 13 C9 19
Blake Gro. 7 J8 22
Blakeney Gro. 10 J14 38
Blakeney Rd. 10 J14 38
Blayds Yd. 1 J11 30
Blencarn Cl. 14 N9 24
Blencarn Garth, 14 N9 24
Blencarn Lawn, 14
Blencarn Cl. N9 24
Blencarn Path, 14
Brooklands Av. N9 24
Blencarn Rd. 14 N9 24
Blencarn Vw. 14 N9 24
Blenheim Av. 2 H10 29
Blenheim Cres. 2
Blenheim Av. H10 29
Blenheim Gro. 7 H10 29
Blenheim Mt. 7 H10 29
Blenheim Sq. 2 H10 29
Blenheim Ter. 2 H9 21
Blenheim Wk. 2 H10 29
Blezard Ct. 11
Meadow La. J12 30
Blind Lane, BD11 A15 34
Blind Lane, 14 N5 16
Bloomfield Mt. 6
Holborn St. H9 21
Bloomfield Pl. 6
Holborn St. H9 21
Bloomfield Ter. 6
Holborn St. H9 21

Blucher St. 12 G11 29
Blue Hill Cres. 12 E12 28
Blue Hill Gro. 12 E12 28
Blue Hill Lane, 12 E12 28
Blundell Ter. 1
Caledonian Rd. H10 29
Boar Lane, 1 H11 29
Bodley St. 4 F10 28
Bodmin App. 10 H16 42
Bodmin Cl. 10 H17 42
Bodmin Cres. 10
H16 to H17 42
Bodmin Croft, 10 H16 42
Bodmin Gdns. 10 H16 42
Bodmin Gdns. 10 H17 42
Bodmin Garth, 10 H17 42
Bodmin Pl. 10 H17 42
Bodmin Rd. 10 G16 42
Bodmin St. 10 H17 42
Bodmin Sq. 10 H17 42
Bodmin Ter. 10 H17 42
Boggard Lane, 13 C10 27
Boggart Hill, 14 N8 24
Boggart Hill Ct. 14 N8 24
Boggart Hill Cres. 14 N8 24
Boggart Hill Dr. 14 N8 24
Boggart Hill Gdns.
14 N8 24
Boggart Hill Gra. 14 N8 24
Boggart Hill Rd. 14 N8 24
Boggart Hill Towers,
14 N8 24
Boldmere Rd. 15 N11 32
Bold St. 12
Whitehall Rd. G12 29
Bolland's Ct. 6
Pennington St. H9 21
Bolland St. 6 H9 21
Bond St. 1 H11 29
Booth Pl. 6
North West Rd. H9 21
Booth St. 11
Sweet St. H12 29
Borough Vw. 8 K7 22
Borrough Av. 8 K7 22
Borrowdale Ter. 14 N9 24
Bosnia Gro. 12 E11 28
Bosnia Pl. 12
Whingate Rd. E11 28
Bosphorous St. 12 E11 28
Boston Av. 5 D9 19
Bottomley's Bldgs. 6 H9 21
Bottoms, The, 27 C15 35
Boundary Farm Road.
17 H5 13
Boundary St. 7
Roundhay Rd. K9 22
Bower Rd. 10 K13 38
Bower Yd. 10
Woodhouse Hill Rd. K14 38
Bowfell Cl. 14 O9 24
Bowland Cl. 15 N11 32
Bowling Green Ter.
11 H12 29
Bowman Lane, 10 J11 30
Bowness St. 11
Cambrian Rd. H13 37
Bowood Av. 7 H7 21
Bowood Cres. 7 H7 21
Bowood Gro. 7 H7 21
Bow St. 9 K11 30
Boyle St. 12 E11 28
Boyne Pl. 10 J12 30
Boyne St. 10
Holdsworth St. J12 30
Bracken Edge, 8 K8 22
Bracken Hill, 17 J6 14
Brackenwood Dr. 8 K7 22
Brackenwood Grn. 8 K7 22
Bramley Gro. 13 C10 27
Bramley Hill, 5 D9 19
Bramley Mt. 13 C10 27
Bramley Pl. 13 C10 27
Bramley St. 13 C10 27
Bramstan Av. 13 B9 18
Bramstan Gdns. 13 B9 18
Brancepeth Pl. 12 G11 29
Branch Pl. 12 E13 36

Branch Rd. 27   C15 35
Branch Rd. 12   E13 36
Branch St. 12
   Branch Rd.   E13 36
Branch Church St. 10
   Waterloo Rd.   K13 38
Brander App. 9   M10 31
Brander Cl. 9   M10 31
Brander Dr. 9   M10 31
Brander Gro. 9   M10 31
Brander Mt. 9   M10 31
Brander Rd. 9   M10 31
Brander St. 9   M10 31
Brandling Pl. 10
   Longwood Pl.   J13 38
Brandon Gro. 7   J8 22
Brandon La. 17
     M3 to N4   9
Brandon Rd. 3   H10 29
Brandon St. 12   G11 29
Brandon Ter. 17   L4 15
Brandon Way, 7   J8 22
Branksome Pl. 6
   Queens Rd.   G10 29
Branksome Ter. 6
   Queens Rd.   G10 29
Brantford St. 7   J8 22
Brathay Gdns. 14   N9 24
Brayton App. 14   P8 25
Brayton Cl. 14   P8 25
Brayton Garth, 14   P8 25
Brayton Grn. 14   P8 25
Brayton Gro. 14   P8 25
Brayton Pl. 14   P8 25
Brayton Sq. 14   P8 25
Brayton Ter. 14   P8 25
Brayton Wk. 14   P8 25
Breary Av. 18   C6 11
Breary Lane, 16   C1   4
Breary Rise, 16   D2   5
Breary Ter. 18   C6 11
Breary Wk. 18   C6 11
Brecon App. 9   M10 31
Brecon Ct. 9
   Brecon Approach   M10 31
Brecon Rise, 9
   Brecon Approach   M10 31
Brentwood Gro. 12
   Brooklyn Ter.   F11 28
Brian Cres. 15   O9 24
Brian Pl. 15   O9 24
Brian Vw. 15   O9 24
Briarsdale Ct. 8   M9 23
Briarsdale Croft, 8   M9 23
Briarsdale Garth, 8   L9 23
Briarsdale Hts. 9   M9 23
Brick Row, 6
   Green Rd.   G7 21
Brick Mill Rd. 28   A12 26
Bridge End, 1   J11 30
Bridge Rd. 13   A8 18
Bridge Rd. 5   E9 20
Bridge Rd. 11   H12 29
Bridge St. 2   J11 30
Bridge Ter. 11
   Bay Horse La.   O5 16
Bridge View, 9
   S. Accommodation Rd. J12 30
Bridge View. 13   A8 18
Bridgefield Pl. 9
   S. Accommodation Rd. J12 30
Bridgewater Pl. 9   K12 30
Bridgewater Rd. 9   K12 30
Bridgewater Ter. 9   K12 30
Bridle Path, 15   O10 33
Bridle Path Cres. 15   O10 33
Bridle Path Rd. 15   O10 33
Bridle Path Sq. 15   O10 33
Bridle Path Wk. 15   O10 33
Briggate, 1   J11 30
Brighton Gro. 13   C10 27
Bright St. Stan.   A10 26
Brignall Croft, 9   K10 30
Brignall Garth, 9   K10 30
Brignall Way, 9   K10 30
Bristol Rd. 7   J10 30
Bristol St. 7   J10 30
Britannia Ct. 13   B11 26

Britannia St. 28   A10 26
Britannia St. 1
   Wellington St.   G11 29
Broadgate Av. 18   C6 11
Broadgate Cres. 18   C7 19
Broadgate Dr. 18   C6 11
Broadgate La. 18   C6 11
Broadgate Wk. 18   C7 19
Broad Lane,
   13   B10 to D9 26
Broadlea Av. 13   D9 19
Broadlea Cl. 13   D9 19
Broadlea Cres. 13   D9 19
Broadlea Gdns. 13   D9 19
Broadlea Gro. 13   D9 19
Broadlea Hill, 13   D9 19
Broadlea Mt. 13   D9 19
Broadlea Pl. 13   D9 19
Broadlea Rd. 13   D9 19
Broadlea St. 13   D9 19
Broadlea Ter. 13   D9 19
Broadlea Vw. 13   C9 19
Broadway, 13 & 18   A8 18
Broadway, 15   N11 32
Broadway Dr. 18   B7 18
Broadway Pl. 11
   Malvern Rd.   H13 37
Broadway St. 11
   Malvern Rd.   H13 37
Brompton Gro. 11
   Stratford Ter.   H13 37
Brompton Mt. 11
   Stratford Ter.   H13 37
Brompton Row, 11
   Stratford Ter.   H13 37
Brompton Ter. 11
   Stratford Ter.   H13 37
Brompton Vw. 11
   Stratford Ter.   H13 37
Brookdale Ter. 11
   Malvern Ter.   H13 37
Brookfield Av. 8   K9 22
Brookfield Pl. 6
   Brookfield Rd.   G8 21
Brookfield Rd. 6   G8 21
Brookfield Ter. 6
   Monk Bridge Rd.   G8 21
Brookfoot, 18   C6 11
Brookhill Av. 17   K5 14
Brookhill Cl. 17   K5 14
Brookhill Cres. 17   K5 14
Brookhill Dr. 17   K5 14
Brookhill Gro. 17   K5 14
Brooklands Av. 14   N9 24
Brooklands Cl. 14   N9 24
Brooklands Cres. 14   N9 24
Brooklands Dr. 14   N9 24
Brooklands Garth, 14 N9 24
Brooklands La. 14
     N9 to O8 24
Brooklands Pl. 14   N9 24
Brooklands Vw. 14   N9 24
Brooklyn Av. 12   F11 28
Brooklyn Pl. 12   F11 28
Brooklyn St. 12   F11 28
Brooklyn Ter. 12   F11 28
Broom Cres. 10   K15 38
Broomfield, 16   E5 12
Broomfield Cres. 6   F9 20
Broomfield Pl. 6   F9 20
Broomfield Rd. 6   F9 20
Broomfield St. 6   F9 20
Broomfield Ter. 6   F9 20
Broomfield Vw. 6
   Chapel La.   F9 20
Broom Gdns. 10   K15 38
Broom Garth, 10   K15 38
Broom Gro. 10   K16 43
Broomhill Av. 17   J6 14
Broomhill Cres. 17   J6 14
Broomhill Dr. 17   J6 14
Broomhill Rd. 9   L10 33
Broom Mt. 10   K16 48
Broom Pl. 10   K.5 33
Broom Rd. 10   K16 48
Broom Ter. 10   K15 31
Broughton Av. 11   L10 31
Broughton Ter. 28   A11 26

Broughton Ter. 9   L10 31
Brown Av. 11   G13 37
Brownberrie Av.
   18   C5 11
Brownberrie Cres.
   18   B5 10
Brownberrie Dr.
   18   C5 11
Brownberrie Gdns.
   18   B5 10
Brownberrie La.
   18   A5 10
Brownberrie Wk.
   18   C5 11
Brown Hill Av. 9   L10 31
Brown Hill Cres. 9   L10 31
Brown Hill Ter. 9   L10 31
Brown La. 12 and 11 G12 29
Brown Pl. 11   G13 37
Brown Rd. 11   G12 29
Brown's Pl. 10   K13 38
Brown's Sq. 7
   Skinner La.   J10 30
Brown's Ter. 13   C10 27
Bruce Gdns. 12   G11 29
Bruce Lawn, 12   G11 29
Brudenell Av. 6   G9 21
Brudenell Gro. 6   G9 21
Brudenell Mt. 6   G9 21
Brudenell Rd. 6   G9 21
Brudenell St. 6   G9 21
Brudenell Vw. 6   G9 21
Brunel Pl. 12   G11 29
Brunel St. 12   G11 29
Brunswick Pl. 2   J10 30
Brunswick Rd. 28   A11 26
Brunswick Ter. 2   J10 30
Brussels St. 9   J11 30
Buckingham Av. 6   G9 21
Buckingham Dr. 6   G9 12
Buckingham Gro. 6   G9 21
Buckingham Mt. 6   G9 21
Buckingham Rd. 6   G9 21
Buckley Av. 11   H13 37
Buckstone Av. 17   G5 13
Buckstone Cl. 17   H5 13
Buckstone Cres. 17   G5 13
Buckstone Dr. 17   G5 13
Buckstone Gdns. 17   H5 13
Buckstone Grn. 17   G5 13
Buckstone Gro. 17   G5 13
Buckstone Mt. 17   G5 13
Buckstone Oval, 17   G5 13
Buckstone Rise, 17   G5 13
Buckstone Rd. 17   G5 13
Buckstone Vw. 17   G5 13
Buckton Pl. 11   H13 37
Buckton Rd. 11   H13 37
Buckton St. 11   H13 37
Buckton Ter. 11
   Buckton Rd.   H13 37
Bude Rd. 11   H14 37
Bulgaria St. 12   E11 28
Bull and Bell Yd. 1
   Briggate   J11 30
Buller Cres. 9   M10 31
Bullerthorpe La.
   15 and 26   Q14 to Q12 41
Bullough La. 26   N15 40
Bulmer St. 7
   Meanwood Rd.   G8 21
Burchett Gro. 6   H9 21
Burchett Pl. 6   H9 21
Burchett Ter. 6
   Hartley Av.   H9 21
Burley Grange Rd. 4 F10 28
Burley Hill Cres. 4   E9 20
Burley Hill Dr. 4   E9 20
Burley La. 18   B7 18
Burley Lodge Pl. 6
   Burley Lodge Rd.   G10 29
Burley Lodge Rd. 6 G10 29
Burley Lodge Ter. 6 G10 29
Burley Pl. 4   F10 28
Burley Rd. 4 and 3   F10 28
Burley St. 3   G11 29
Burley Wood Cres. 4 E9 20
Burley Wood La. 4   F9 20

49

Burley Wood Mt. 4   E9 20
Burley Wood Vw. 4   F9 20
Burlington Pl. 11
  Tempest Rd.   H13 37
Burlington Rd. 11   H14 37
Burmantofts St. 9   K11 30
Burnsall Cl. 12   E11 28
Burnsall Ct. 12   E11 28
Burnsall Gdns. 12   E11 28
Burnsall Gro. 12   E11 28
Burnt Hills Rd. 12   C14 35
Burton Av. 11   J13 38
Burton Cres. 6   F8 20
Burton Gro. 11   J13 38
Burton Rd. 11   J13 38
Burton Row, 11   J13 38
Burton St. 11   J13 38
Burton Ter. 11   J13 38
Burton Vw. 11
  Tunstall Rd.   J13 38
Burton Way, 9   L10 31
Bushire St. 12
  Armley Rd.   F11 28
Buslingthorpe Grn. 7   J9 22
Buslingthorpe La. 7   J9 22
Butcher Hill, 18
  and 16   D7 19
Butcher St. 11
  Water La.   H12 29
Butterbowl Dr. 12   D12 27
Butterbowl Gdns. 12   D12 27
Butterbowl Garth 12   D12 27
Butterbowl Gro. 12   D12 27
Butterbowl Lawn, 12   D12 27
Butterbowl Mt. 12   D12 27
Butterbowl Rd. 12   D12 27
Butterfield St. 9   K11 30
Butterley St. 10   J12 30
Butt Gdns. 12   C12 27
Butt Grn. 12   C12 27
Butt La. 12   C12 27
Butts Ct. 1   J11 30
Butts Gro. 12
  Whiteley St.   F11 28
Butts Mt. 12   F11 28
Butts Pl. 12
  Masham St.   F11 28
Butts St. 12   F11 28
Butts Ter. 12
  Masham St.   F11 28
Butts Vw. 12
  Whiteley St.   F11 28
Byron St. 28   A12 26
Byron St. 2   J10 30

Cabbage Hill, 12
  Upper Wortley Rd.   E11 28
Cad Beeston, 11
  Cemetery Rd.   H13 37
Calgary Pl. 7   J8 22
Call Lane, 1   J11 30
Calls, The, 2   J11 30
Calverley Av. 13   B9 18
Calverley Ct. 13   B9 18
Calverley Dr. 13   B9 18
Calverley Gdns. 13   B9 18
Calverley Garth, 13   B9 18
Calverley Gro. 13   B9 18
Calverley La. 13   B9 18
Calverley La. 18   A7 18
Calverley St. 1   H10 29
Calverley Ter. 13   B9 18
Camberley St. 11
  Dewsbury Rd.   G15 37
Cambrian Rd. 11   H13 37
Cambrian St. 11   H13 37
Cambrian Ter. 11
  St. Luke's Rd.   H13 37
Cambridge Gdns. 13   B9 18
Cambridge Gardens Est.
  Leeds & Bradford Rd. A9 18
Cambridge Rd. 7   H9 21
Camden Ter. 2
  Woodhouse La.   G9 21
Camp Rd. 16   C2 5

Canaan Sq. 10
  Low Rd.   K13 38
Canal Pl. 12
  Armley Rd.   F11 28
Canal Rd. 13   A8 18
Canal Rd. 12   F11 28
Canal St. 12   G11 29
Canal Wharf, 11
  Water La.   H12 29
Cancel St. 10   J12 30
Canning St. 11   H12 29
Canonbury Gro. 11
  Wesley St.   G13 37
Canonbury Mt. 11
  Wesley St.   G13 37
Canonbury St. 11
  Wesley St.   G13 37
Canonbury Ter. 11
  Wesley St.   G13 37
Canterbury Dr. 6   F9 20
Canterbury Rd. 6   F9 20
Carberry Pl. 6
  Burley Lodge Rd.   G10 29
Carberry Rd. 6   G10 29
Carberry Ter. 6
  Burley Lodge Rd.   G10 29
Carden Av. 15   N11 32
Cardigan La. 4   F10 28
Cardigan La. 6   F9 20
Cardigan Pl. 6   F8 20
Cardigan Rd. 6 F9 to G10 20
Cardigan Row, 4   F10 28
Cardigan Ter. 4   F10 28
Cardigan Vw. 4   F10 28
Cardinal Av. 11   G15 37
Cardinal Cres. 11   G15 37
Cardinal Gdns. 11   G15 37
Cardinal Garth, 11
  Cardinal Sq.   G14 37
Cardinal Gro. 11   G15 37
Cardinal Rd. 11   G15 37
Cardinal Sq. 11   G14 37
Cardinal Wk. 11   G14 37
Carlisle Rd. 7
  Albert Gro.   J10 30
Carlisle Rd. 10   J12 30
Carlisle Rd. 28   A12 26
Carlton Av. 28   A11 26
Carlton Carr, 7   J10 30
Carlton Cl. 7   J10 30
Carlton Croft, 7   J10 30
Carlton Gdns. 7   J10 30
Carlton Gate, 7   J10 30
Carlton Garth, 7   J10 30
Carlton Par. 7   J10 30
Carlton Rise, 7   J10 30
Carlton Rise, 28   A11 26
Carlton Row, 12   E11 28
Carlton St. 2 and 7   H10 29
Carlton Towers, 7   J10 30
Car Moor Side,
  10 and 11   J13 38
Carr Bridge Av. 16   C5 11
Carr Bridge Dr. 16   C5 11
Carr Bridge Vw. 16   C5 11
Carr Crofts, 12   E11 28
Carr Crofts Ter. 12
  Tong Rd.   E11 28
Carr Cross St. 7   J9 22
Carrholm Cres. 7   H7 21
Carrholm Dr. 7   H7 21
Carrholm Gro. 7   H7 21
Carrholm Mt. 7   H7 21
Carrholm Rd. 7   H7 21
Carrholm Vw. 7   H7 21
Carriage Dr. The, 8   M7 23
Carr La. 14   O5 16
Carr Manor Av. 17   H7 21
Carr Manor Cres. 17   H6 13
Carr Manor Croft, 17   H7 21
Carr Manor Dr. 17   H7 21
Carr Manor Gdns. 17   H7 21
Carr Manor Garth, 17 H6 13
Carr Manor Gro. 17   H7 21
Carr Manor Mt. 17   H7 21
Carr Manor Par. 17   H7 21
Carr Manor Pl. 7   H7 21
Carr Manor Rd. 17   H6 21

Carr Manor Vw.
  17   H6 21
Carr Manor Wk. 7   H7 21
Carr Ter. 10
  Balm Rd.   K14 38
Carter Av. 15   P11 33
Carter La. 15   P11 33
Carter Mt. 15   P11 33
Carter Ter. 15   P10 33
Carter Ter. 11
  Town St.   G14 37
Cartmell Dr. 15   N11 32
Casterton Cl. 14   O9 24
Casterton Gdns. 14   O9 24
Castle Grove Av. 6   F8 20
Castle Grove Dr. 6   F8 20
Castle Ings Cl. 12   C13 35
Castle Ings Dr. 12   C13 35
Castle Ings Gdns. 12 C13 35
Castle St. 1   H11 29
Castleton Ter. 12
  Armley Rd.   F11 28
Castle Vw. 17   H6 11
Cathcart St. 6   H9 23
Catherine Gro. 11
  Lodge La.   H13 37
Causeway La. 16
  Westwood La.   F6 12
Cautley Pl. 13   C10 27
Cautley Rd. 9   K12 30
Cautley St. 13   C10 27
Cavalier St. 9   K12 30
Cavendish Pl. 28   A10 26
Cavendish Rise, 28   B11 26
Cavendish St. 3   G11 29
Cave St. 11   J12 30
Cawood Yd. 9   J11 30
Caythorpe Rd. 16   E7 20
Cecil Gro. 12   F11 28
Cecil Mt. 12   F11 28
Cecil Rd. 12   F11 28
Cecil St. 12   F11 28
Cedar Av. 12   E11 28
Cedar Gro. 12
  Carr Crofts   E11 28
Cedar Mt. 12   E11 28
Cedar Pl. 12   E11 28
Cedar Rd. 12   E11 28
Cedar St. 12   E11 28
Cedar Ter. 12   E11 28
Cedars, The, 16   D2 5
Cemetery Pl. 2
  Woodhouse La.   G9 21
Cemetery Rd. 28   A11 26
Cemetery Rd. 11   H13 37
Centenary St. 1   H11 29
Central Rd. 1   J11 30
Central St. 1
  St. Paul's St.   H11 29
City Sq. 1   H11 29
Chadwick St. 10   J12 30
Chalfont Rd. 16   E7 20
Chambers St. 10   K12 30
Chancellor Pl. 6
  Speedwell St.   H9 21
Chancellor St. 6
  Ridge Rd.   H9 21
Chandos Av. 8   K7 22
Chandos Fold, 8   K7 22
Chandos Gdns. 8   K7 22
Chandos Garth, 8   K7 22
Chandos Gro. 8   K7 22
Chandos Pl. 8   K7 22
Chandos Ter. 8   K7 22
Chandos Wk. 8   K7 22
Chantrell Gro. 9
  York Rd.   J11 30
Chantrell Pl. 9
  Chantrell St.   K11 30
Chapel Ct. 11
  Marshall St.   H12 29
Chapel Hill, 10   J16 43
Chapel La. 12   F11 28
Chapel La. 12   C13 35
Chapel La. 6   F9 20
Chapel La. 15   P12 33
Chapel Pl. 6
  North La.   F9 20

50

Street	Grid	Page
Chapel Rd. 7	J8	22
Chapel Row, 11		
Marshall St.	H12	29
Chapel Sq. 6		
Chapel St.	F8	20
Chapel St. 15	O11	32
Chapel St. 6	F8	20
Chapel St. 28	A10	26
Chapeltown Rd. 7	J10	30
Chapel Vw. 18	C7	19
Chapel Yd. 15		
Chapel St.	O11	32
Chariot St. 1	H11	29
Charles Av. 9	K12	30
Charles Gdns. 11	H12	29
Charles St. 18	B7	18
Charlton Gro. 9	K11	30
Charlton Mt. 9	K11	30
Charlton Pl. 9	K11	30
Charlton Rd. 9	K11	30
Charlton St. 9	L11	31
Charlville Gdns. 17	N5	16
Charmouth Pl. 11		
Charmouth St.	H12	29
Charmouth St. 11	H12	29
Charmouth Ter. 11	H13	37
Charville Gdns. 17	N5	16
Chatswood Av. 11	G15	37
Chatswood Cres. 11	G15	37
Chatswood Dr. 11	G14	37
Chatsworth Rd. 8	L9	23
Chatsworth St. 12		
Leamington Ter.	F12	28
Chatsworth Ter. 12		
Leamington Ter.	F12	28
Chaucer Av. 28	A12	26
Chaucer Gdns. 28	A12	26
Checker Row, 28		
Richardshaw La.	A10	26
Chelwood Av. 8	K5	14
Chelwood Cres. 8	K6	14
Chelwood Dr. 8	K5	14
Chelwood Gro. 8	K5	14
Chelwood Mt. 8	K5	14
Chelwood Pl. 8	K5	14
Chenies Cl. 14	N10	32
Cherry Pl. 9	K10	30
Cherry Ct. 9	K10	30
Cherry Row, 9	K10	30
Chesney Av. 10	J13	38
Chestnut Av. 6	G9	21
Chestnut Av. 15	P10	33
Chetwynd St. 11	H12	29
Chetwynd Ter. 11		
Meadow Rd.	H12	29
Chichester St. 12	F11	28
Chiltern Gro. 8	L8	23
Chiswick St. 6		
Burley Lodge Rd.	G10	29
Chiswick Ter. 6		
Burley Lodge Rd.	G10	29
Christchurch Av. 12		
Stanningley Rd.	D10	27
Christchurch Mt. 12		
Stanningley Rd.	D10	27
Christchurch Par. 12		
Moorfield Rd.	E11	28
Christchurch Rd. 12	E11	28
Christchurch Ter. 12		
Moorfield Rd.	E11	28
Christchurch Vw. 12		
Stanningley Rd.	D10	27
Christopher Rd. 6	H9	21
Church Av. 27	C15	35
Church Av. 18	B6	10
Church Av. 6	G7	21
Church Cl. 14	O9	24
Church Cres. 18	B6	10
Church Cres. 17	J5	14
Churchfield Pl. 5		
Church St.	E9	20
Churchfield Pl. 6		
Holborn St.	H9	21
Churchfield St. 6		
Institution St.	H9	21
Churchfield Ter. 6		
Holborn St.	H9	21
Church Gdns. 17	J5	14
Church Gro. 18	B6	10
Church Hill Mt. 28		
Leeds & Bradford Rd.	A10	18
Church Hill Pl. 28		
Leeds & Bradford Rd.	A10	18
Church Hill St. 28		
Leeds & Bradford Rd.	A10	18
Churchill Gdns. 2		
Blenheim Walk	H10	29
Church Hill Ter. 28		
Leeds & Bradford Rd.	A10	18
Church La. 28	A11	26
Church La. 26	P15	41
Church La. 7	J8	22
Church La. 6	G7	21
Church La. 5	E9	20
Church La. 16	F5	12
Church La. 2	J11	30
Church La. 15	P10	33
Church Mt. 18	B6	10
Church Rd. 18	B6	10
Church Rd. 9	K11	30
Church Rd. 12	F11	28
Church Row, 2		
Kirkgate	J11	30
Church St. 27	B15	34
Church St. 5	E9	20
Church St. 10	J13	38
Church Ter. 7	J7	22
Church Vw. 18	B6	10
Church Vw. 5	E9	20
Church Wk. 18	B6	10
Church Wood Av. 16	F8	20
Church Wood Mt. 16	F7	20
Church Wood Rd. 16	F8	20
Churwell, 11	G15	37
Churwell Bar,		
Elland Rd.	G13	37
Clapgate La. 10	K16	43
Clapham Dene Row, 15	O10	32
Claremont Av. 3	H10	29
Claremont Cres. 6	G8	21
Claremont Dr. 6	G8	21
Claremont Pl. 12	E11	28
Claremont Rd. 6	G8	21
Claremont St. 12	E11	28
Claremont Ter. 12	E11	28
Claremount Gro. 3		
Claremount Av.	H10	29
Claremount Vw. 3		
Claremount Av.	H10	29
Claremount Yd. 10		
Waterloo Rd.	K13	38
Clarence Dr. 18	B7	18
Clarence Gdns. 18	B7	18
Clarence Gro. 18	B7	18
Clarence Rd. 18	B7	18
Clarence Rd. 10	J12	30
Clarence St. 7	J9	22
Clarence St. 13	C10	27
Clarendon Rd. 2	H10	29
Clarendon Ter. 28	A12	26
Clark Av. 9	K11	30
Clark Ct. 9	K11	30
Clark La. 9	K12	30
Clark Mt. 9		
Pontefract La.	K11	30
Clark Pl. 11	H12	29
Clark Rd. 9	K11	30
Clarkson's Bldgs. 6		
Woodhouse St.	H9	21
Clarkson Vw. 6	H9	21
Clark St. 10	K12	30
Clark Ter. 9	K11	30
Clark Vw. 9	K12	30
Claro Av. 7	H9	21
Claro Rd. 7	H9	21
Claro St. 7		
Cambridge Rd.	H9	21
Claro Ter. 7		
Cambridge Rd.	H9	21
Claro Vw. 7	H9	21
Clayfield Pl. 7	J9	22
Clayfield St. 7		
Oxford Rd.	H9	21
Clayfield Ter. 7		
Oxford Rd.	H9	21
Clay Pit Lane, 2	J10	30
Clay Pit St. 7	J10	30
Clayton Ct. 16	D7	19
Clayton Gra. 16	D7	19
Clayton Wood Cl. 16	D6	11
Clayton Woods Rise, 16	D6	11
Clayton Wood Rd. 16	D6	11
Cleopatra Pl. 13	C10	27
Cleopatra St. 13	C10	27
Cleveleys Av. 11	G13	37
Cleveleys Mt. 11		
Cleveleys Rd.	G13	37
Cleveleys Rd. 11	G13	37
Cleveleys St. 11		
Cleveleys Rd.	G13	37
Cleveleys Ter. 11		
Cleveleys Rd.	G13	37
Cliffdale Rd. 7	H9	22
Cliff Lane, 6	G9	21
Cliff Mount, 6	H9	22
Cliff Mount Ter. 6		
Cliff Mount	H9	21
Cliff Pit St. 6	H9	21
Cliff Pl. 6	H9	21
Cliff Rd. 6	H9	21
Cliff Ter. 18	C7	19
Cliff Ter. 6	H9	21
Clifton Av. 9	L10	31
Clifton Dr. 28	A11	26
Clifton Gro. 9	L10	31
Clifton Hill, 28	A11	26
Clifton Mt. 9	L10	31
Clifton Rd. 28	A11	26
Clifton Ter. 9	L10	31
Clipston Av. 6	G8	21
Clipston St. 6	G8	21
Clive St. 11	H12	29
Close, The, 16	D2	5
Close, The, 18	A5	10
Close, The, 17	H4	13
Close, The, 9	K11	30
Cloth Hall St. 2		
Call La.	J11	30
Clothier St. 10		
Waterloo Rd.	K13	38
Clovelly Gro. 11		
Rowland Rd.	H13	37
Clovelly Pl. 11		
Rowland Rd.	H13	37
Clovelly Row, 11		
Rowland Rd.	H13	37
Clovelly Ter. 11	H13	37
Clowes St. 11		
Jack La.	H12	29
Club Lane, 13	A8	18
Club Row, 5	E9	20
Club Row, 28		
Richardshaw La.	A10	26
Club Ter. 5	E9	20
Clyde App. 12	G11	29
Clyde Ct. 12	G12	29
Clyde Gdns. 12	G12	29
Clyde Gra. 12		
Clyde Vw.	G12	29
Clyde Vw. 12	G12	29
Clyde Wk. 12	G11	29
Coal Hill La. 28 and 13	A9	18
Coal Rd. 14	O5 to O7	05
Coal Staith Rd. 10		
Hunslet La.	J12	30
Cobden Av. 12	D13	35
Cobden Gro. 12	E13	36
Cobden Pl. 12	D13	35
Cobden Rd. 12	D13	35
Cobden St. 12	D13	35
Cobden Ter. 12	D13	35
Cockshott Cl. 12	D10	27
Cockshott Dr. 12	D10	27
Cockshott La. 12	D10	27
Coggill St. 10	L14	39
Colby Rise, 15	N11	32
Coldcall Pl. 6	H9	21
Coldcall St. 6	H9	21
Coldcall Ter. 6		
Woodhouse St.	H9	21
Coldcotes Av. 9	L10	31

Coldcotes Circus, 9 M10 31
Coldcotes Cl. 9 M10 31
Coldcotes Cres. 9 M10 31
Coldcotes Dr. 9 M10 31
Coldcotes Garth, 9 M10 31
Coldcotes Gro. 9 M10 31
Coldcotes Vw. 9 M10 31
Coldcotes Wk. 9 M10 31
Coldwell Rd.
  15 O10 to P10 32
Coldwell Sq. 15 O10 32
Coleman St. 12 G12 29
Colemore Gro. 12 F12 28
Colemore Pl. 12
  Oldfield La. F12 28
Colemore Rd. 12 F12 28
Colemore St. 12 F12 28
Colenso Gro. 11 G13 37
Colenso Mt. 11 G13 37
Colenso Pl. 11 G13 37
Colenso Rd. 11 G13 37
Colenso Ter. 11 G13 37
College Rd. 27 C15 35
Colliers La. 17 N5 16
Collin Rd. 14 N10 32
Colton Rd. 12 F11 28
Colton Rd. 15 P11 33
Colton Rd. E.15 Q12 33
Colton St. 12 F11 28
Colville Ter. 11 H13 37
Colwyn Av. 11
  Colwyn Rd. H14 37
Colwyn Mt. 11
  Colwyn Rd. H14 37
Colwyn Rd. 11 H14 37
Colwyn Ter. 11
  Colwyn Rd. H14 37
Colwyn Vw. 11 H14 37
Commercial Ct. 1
  Briggate J11 30
Commercial Rd. 5 E9 20
Commercial St. 1 J11 30
Commercial Ter. 11
  Marshall St. H12 29
Compton Av. 9 L10 31
Compton Cres. 9 L10 31
Compton Gro. 9
  Compton Av. L10 31
Compton Mt. 9
  Compton Av. L10 31
Compton Pl. 9 L10 31
Compton Rd. 9 L10 31
Compton Row, 9 L10 31
Compton St. 9
  Compton Av. L10 31
Compton Ter. 9 L10 31
Compton Vw. 9 L10 31
Concordia St. 1 J11 30
Concord St. 2 J10 30
Conference Pl. 12 E11 28
Conference Rd. 12 E11 28
Conference Ter. 12
  Conference Rd. E11 28
Congress Mt. 12 E11 28
Congress Pl. 6
  Institution St. H9 21
Congress St. 12 E11 28
Coniston Av. 6 G8 21
Coniston Mt. 28
  Leeds & Bradford Rd. A10 18
Coniston Pl. 28
  Leeds & Bradford Rd. A10 18
Coniston St. 28
  Leeds & Bradford Rd. A10 18
Coniston Ter. 28
  Leeds & Bradford Rd. A10 18
Coniston Way, 26 P15 41
Consort St. 3 G10 29
Consort Ter. 3 G10 29
Constance St. 6
  Lucas St. H9 21
Conway Av. 8 K9 22
Conway Dr. 8 K9 22
Conway Gro. 8 K9 22
Conway Mt. 8 K9 22
Conway Pl. 8 K9 22
Conway Rd. 8 K9 22
Conway St. 8 K9 22

Conway Ter. 8 K9 22
Conway Vw. 8 K9 22
Conyer's Fold, 11
  Wortley La. G12 29
Conyer's Yd. 11
  Wortley La. G12 29
Cookridge Av. 16 C4 11
Cookridge Dr. 16 C4 11
Cookridge Gro. 16 D4 11
Cookridge La. 16 C3 5
Cookridge St. 2 H11 29
Cookson St. 9
  Ellerby Rd. K11 30
Cooper Hill, 28 A12 26
Cooper's Ct. 10
  Kendell St. J11 30
Copgrove Rd. 8 L8 23
Copley Hill, 12 G12 29
Copley Yd. 12
  Tong Rd. E11 28
Copperfield Av. 9 K12 30
Copperfield Ct. 9 K12 30
Copperfield Gro. 9 K12 30
Copperfield Mt. 9
  Cross Green La. K12 30
Copperfield Pl. 9 K12 30
Copperfield Ter. 9 L12 31
Copperfield Vw. 9 K12 30
Coppice Way, 8 L8 23
Coppy Lane, 13 C9 19
Corn Exchange 1
  Call La. J11 30
Corn Mill Fold. 18 D7 19
Coronation Par. 15 N12 32
Coronation Pl. 2
  Albion St. J11 30
Coronation St. 2
  Woodhouse La. H9 21
Cottage Rd. 6 F8 20
Cottage St. 9 K11 30
Cotterdale Vw. 15 N12 32
Cottingley App. 11 F14 36
Cottingley Clo. 11 F14 36
Cottingley Cres. 11 F14 36
Cottingley Dr. 11 F14 36
Cottingley Gdns. 11 F14 36
Cottingley Garth,
  11 F14 36
Cottingley Grn. 11 F14 36
Cottingley Gro. 11 F14 36
Cottingley Pl. 11 F14 36
Cottingley Wk. 11 F14 36
Cotton Mill Row, 10
  Arthington Av. J13 38
Cotton St. 9 J11 30
County Arcade, 1
  Briggate J11 30
Coupland Pl. 11
  Coupland St. H13 37
Coupland St. 11 H13 37
Cow Close Gro. 12 E13 36
Cow Close Rd. 12 D12 27
Cowley Rd. 13 A8 18
Cowper Av. 9 L10 31
Cowper Gro. 8 L9 23
Cowper Mt. 9 L10 31
Cowper Rd. 9 L10 31
Cowper St. 7 J9 22
Cowper Ter. 9 L10 31
Crab Lane, 12 F11 28
Cragg Av. 18 B7 18
Cragg Hill, 18 C7 19
Cragg Rd. 18 C7 19
Cragg Ter. 18 B7 18
Cragg Wood Ter.
  18 C7 19
Crag Hill Vw. 16 D4 11
Cragside Cl. 5 D7 19
Cragside Cres. 5 D7 19
Cragside Gdns. 5 D7 19
Cragside Gro. 5 C8 19
Cragside Mt. 5 D7 19
Cragside Pl. 5 D7 19
Cragside Wk. 5 C8 19
Craighill Av. 16 D4 11
Crampton's Bldgs. 10
  St. Helen's St. J12 30
Cranbrook Av. 11 H13 37

Cranbrook Pl. 11
  Dewsbury Rd. H15 37
Cranmer Bank, 17 H5 13
Cranmer Cl. 17 H5 13
Cranmer Gdns. 17 H5 13
Cranmer Rise, 17 H5 13
Cranmer Rd. 17 H5 13
Cranmore Cres. 10 K16 43
Cranmere Dr. 10 K16 43
Cranmore La. 10 K16 43
Cranmore Rise, 10 K16 43
Cranmore Rd. 10 K16 43
Craven Gate, 11
  Moor Cres. J13 38
Craven Rd. 6 H9 21
Craven St. 6 H9 21
Crawshaw Av. 28 A11 26
Crawshaw Gdns.
  28 A11 26
Crawshaw Hill, 28 A11 26
Crawshaw Pk. 28 A11 26
Crawshaw Rise, 28 A11 26
Crawshaw Rd. 28 A11 26
Crescent Av. 26 N15 40
Crescent Gdns. 17 J5 14
Crescent Gra. 11 J13 38
Crescent, The, 15 O11 32
Crescent, The, 28 A11 26
Crescent, The, 13 C9 19
Crescent, The, 18 A6 10
Crescent, The, 17 G4 13
Crescent, The, 16 E5 12
Crescent Towers, 11
  J13 38
Creskeld Cres. 16 D1 5
Creskeld Dr. 16 D1 5
Creskeld La. 16 D2 5
Creskell Gro. 11
  Elland Rd. H12 29
Creskell Pl. 11
  Elland Rd. H12 29
Crest, The, 14 N8 24
Cricklegate, 15 O11 32
Cricketer's Pl. 12 F11 28
Crimbles Rd. 28 A11 26
Crimbles Ter. 28 A11 26
Croftdale Gro. 15 P10 33
Crofton La. 17 O5 16
Crofton Rise, 17 O5 16
Crofton Ter. 17 O5 16
Croft, The, 15 O10 32
Cromer Rd. 2
  Virginia Rd. H10 29
Crompton Ter. 12
  Armley Rd. F11 28
Cromwell Mt. 9 K10 30
Cromwell St. 9 K11 30
Croppergate, 1
  West St. G11 29
Crosby Pl. 11 H12 29
Crosby Rd. 11 H13 37
Crosby St. 11 G12 29
Crosby Ter. 11 H12 29
Crosby Vw. 11
  Crosby St. G12 29
Crosland St. 11
  Bridge Rd. H12 29
Cross Alma St. 9 K10 30
Cross Arcade, 1
  Edward St. J11 30
Cross Aston Gro. 13
  D10 27
Cross Av. 26 N15 40
Cross Aysgarth Mt. 9
  Pontefract La. K11 30
Cross Bank St. 1
  Bank St. J11 30
Cross Barrack St. 7
  Barrack St. J10 30
Cross Barstow St. 11 J12 30
Cross Bath Rd. 13 C10 27
Cross Belgrave St. 2 J11 30
Cross Bentley La. 6 G8 21
Cross Cardigan Mt. 4 F10 28
Cross Chancellor
  St. 6 H9 21
Cross Chapel St. 6
  Chapel St. F8 20

Cross Cliff Rd. 6 G9 21
Cross Ct. I
Briggate J11 30
Cross Cowper St. 7 J9 22
Cross Dawlish Gro. 9
ivy Gro. L11 31
Cross Disraeli St. 11
Disraeli Ter. H13 37
Crossfield St. 2 H9 21
Cross Flatts Av. 11 H14 37
Cross Flatts Cres. 11 G14 37
Cross Flatts Dr. 11 G13 37
Cross Flatts Gro. 11 G14 37
Cross Flatts Par. 11 G14 37
Cross Flatts Pl. 11 G14 37
Cross Flatts Rd. 11 G14 37
Cross Flatts Row, 11 G14 37
Cross Flatts St. 11 G14 37
Cross Flatts Ter. 11 G14 37
Cross Francis St. 7 J9 22
Cross Gates Av. 15 P9 25
Cross Gates La. 15 O9 24
Cross Gates Rd. 15 O10 32
Cross Glen Rd. 16 F7 20
Cross Granby Ter. 6
Bennett Rd. F8 20
Cross Grasmere
St. 12 F11 28
Cross Green. App. 9 L12 31
Cross Green Av. 9 K12 30
Cross Green. Cl. 9 L12 31
Cross Green Cres. 9 K12 30
Cross Green. Dr. 9 L12 31
Cross Green. Garth. 9 L12 31
Cross Green Gro.
Fewston Av. K12 30
Cross Greenhow Av. 4
Greenhow St. F10 28
Cross Greenhow Pl. 4
Greenhow St. F10 28
Cross Greenhow St. 4
Greenhow St. F10 28
Cross Green La. 15 O11 32
Cross Green La. 9 K12 30
Cross Green. Rise, 9 L12 31
Cross Green. Way, 9 L12 31
Cross Harrison St. I
Harrison St. J11 30
Cross Heath Gro. 11 G13 37
Cross Henley Rd. 13 C10 27
Cross Ingledew Cres. 8
Ingledew Cres. L6 15
Cross Ingram Rd. 11 G12 29
Cross John St. 11
Victoria Rd. H12 29
Cross Jubilee Ter. 6
Melville Rd. H9 21
Cross Kelso St. 2 G10 29
Cross Kiln St. 11
Malvern Rd. H13 37
Crossland Rd. 27 E15 36
Crossland St. 11
Dewsbury Rd. G15 37
Crossland Ter. 11
Dewsbury Rd. G15 37
Cross Lane, 12 E11 28
Cross Lane, 12 D12 27
Cross Lea Farm Rd. 5 D7 19
Cross Longley St. 10
Jack La. H12 29
Cross Louis St. 7 J9 22
Cross Mark St. 2
Mark La. J11 30
Cross Mill St. 9
Mill St. J11 30
Cross Myrtle St. 10 J12 30
Cross Newport St. 10
Joseph St. J13 38
Cross Oldfield Gro. 12
Oldfield Av. F12 28
Cross Oldfield Vw. 12
Oldfield Av. F12 28
Cross Osmondthorpe
La. 9 M11 31
Cross Parkfield St. 11 29
Jack La. H12
Cross Princess St. 11
New Princess St. H12 29

Cross Rd. 18 B7 18
Cross Speedwell St. 6
Speedwell St. H9 21
Cross Spenceley St. 2
Mark La. J11 30
Cross Stamford St. 7 J10 30
Cross St. 15 O11 32
Cross Union St. 2
Union St. J11 30
Cross Valley Dr. 15 O10 32
Cross Wingham St. 7
Roundhay Rd. J10 30
Cross Woodview St. 11
Woodview St. H14 37
Cross York St. 2 J11 30
Croton Rd. 12 G11 29
Croton St. 12 G11 29
Crown Ct. 2
Kirkgate J11 30
Crow Nest La. 11 F14 36
Crown Point Rd. 10 J12 30
Crown St. 2 J11 30
Crowther Pl. 6
Woodhouse St. H9 21
Crowther St. 6 H9 21
Croydon Pl. 11 G12 29
Croydon St. 11 G12 29
Croydon Ter. 11
Croydon St. G12 29
Cudbear St. 10 J12 30
Cudworth's Ct. I
Briggate J11 30
Cumberland Rd. 6 G9 21
Cyprus Mt. 6 H9 21
Cyprus Pl. 6
Speedwell Mt. H9 21
Czar St. 11 H12 29

Daffil Rd. 27 E15 36
Daffil Row, 27 E15 36
Dairy St. 6 H9 21
Daisy Cres. 9
Lincoln Rd. K10 30
Daisyfield, 13 C10 27
Daisy Mt. 9 K10 30
Daisy Vw. 9
Lincoln Rd. K10 30
Dale Rd. Dri. B14 34
Dalmeny Pl. 4
Kirkstall Rd. F10 28
Dalmeny St. 4
Kirkstall Rd. F10 28
Dalton Av. 11 H14 37
Dalton Gro. 11 H14 37
Dalton Rd. 11 H14 37
Danby Wk. 9 K11 30
Danube Gro. 12
Gelderd Rd. G12 29
Danube Pl. 12 G12 29
Danube Rd. 12
Gelderd Rd. G12 29
Danube St. 12 G12 29
Danube Ter. 12 G12 29
Danube Vw. 12 G12 29
Darfield Av. 8 L9 23
Darfield Cres. 8 L9 23
Darfield Gro. 8
Harehills Pl. K9 22
Darfield Pl. 8 L9 23
Darfield Rd. 8 L9 23
Darfield St. 8 L9 23
Dargai St. 7 J9 22
Darnley Rd. 16 E7 20
Darnley St. 11
Lady Pit La. H13 37
Dartmouth St. 11 J13 38
David St. 11 H12 29
Davie's Av. 8 K7 22
Dawlish Av. 9 L11 31
Dawlish Cres. 9 L11 31
Dawlish Gro. 9 L11 31
Dawlish Mt. 9 L11 31
Dawlish Pl. 9 L11 31
Dawlish Rd. 9 L11 31
Dawlish Row, 9 L11 31
Dawlish St. 9 L11 31
Dawlish Ter. 9 L11 31
Dawson La. BD4 A14 34

Dawson Rd. 11 H13 37
Dawson St. 28
Sunfield A10 26
Dean Av. 8 L8 23
Dean Ct. 8 L8 23
Deane, The, 15 M12 31
Dean Park Av. BD11 A15 34
Dean Park Dr. BD11 A15 34
Dean St. 3
Kirkstall Rd. F10 28
Deanswood Cl. 17 H5 13
Deanswood Dr. 17 G5 13
Deanswood Gdns. 17 G5 13
Deanswood Garth, 17 H5 13
Deanswood Grn. 17 G5 13
Deanswood Hill, 17 G5 13
Deanswood Pl. 17 H5 13
Deanswood Rise, 17 G5 13
Deanswood Vw. 17 H5 13
De Gray St. 2
Woodhouse La. H9 21
De Lacy Mt. 5 E9 20
Delph La. 6 H9 21
Delph Mt. 6 H9 21
Delph St. 12
Woodhouse St. H9 21
Delph Ter. 6
Woodhouse Cliff H9 21
Denbigh App. 9 M10 31
Denbigh Croft, 9
Denbigh Approach M10 31
Denbigh Heights 9
Denbigh Approach M10 31
Dence Pl. 15 N11 32
Denison Rd. 3 H11 29
Dennil Cres. 15 P9 25
Dennil Rd. 15 P9 25
Dennistead Cres. 6 F8 20
Denton Av. 8 K7 22
Denton Gro. 8 K7 22
Denton Row, 12 E11 28
Dent St. 9 K11 30
Derbyshire La. 16 C3 5
Derbyshire St. 10 K13 38
Derwent Av. 11 H12 29
Derwent Gro. 11
Bath Rd. H12 29
Derwent Vw. 11
Marshall St. H12 29
Derwentwater Gro. 6 F9 20
Derwentwater Ter. 6 F8 20
Detroit Av. 15 P11 33
Detroit Dr. 15 P11 33
Devon Mt. 2
New Camp Rd. H9 21
Devon Rd. 2 H10 29
Devonshire Av. 8 L6 15
Devonshire Cres. 8 L6 15
Devonshire La. 8 L6 15
Dewhurst Pl. 12 F11 28
Dewhurst St. 12 F11 28
Dewsbury Pl. 11
Dewsbury Rd. G15 37
Dewsbury Rd. 11
G15 to J12 37
Dewsbury St. 11
Hunslet Hall Rd. H13 37
Dewsbury Ter. 11
Dewsbury Rd. G15 37
Diadem Dr. 14 N10 32
Dial Lawn, 6
Ramport Rd. H9 21
Dial Row, 6 H9 21
Dial Row, 15 N11 32
Dial St. 9 K12 30
Dial Ter. 9 K12 30
Dib La. 8 M8 23
Dickenson St. 18 C6 11
Dickinson's Ct. I
Boar La. H11 29
Dinsdale Ter. 11
South Ridge St. H13 37
Dixon La. 12 F12 28
Dixon La. Rd. 12 E12 28
Dobson Av. 11 J13 38
Dobson Gro. 11 J13 38
Dobson Pl. 11 J13 38
Dobson Ter. 11 J13 38

Firth Vw. 11 — H14 37
Fir Tree App. 17 — J5 14
Fir Tree Cl. 17 — J5 14
Fir Tree Gdns. 17 — H5 13
Fir Tree Grn. 17 — J5 14
Fir Tree Gro. 17 — J5 14
Fir Tree La. 17 — J5 14
Fir Tree Rise, 17 — J5 14
Fir Tree Vale, 17 — J5 14
Fish St. 1
Kirkgate — J11 30
Fisher's Yd. 11
Meadow La. — J11 30
Fitzarthur St. 12 — F11 28
Fitzroy Dr. 8 — L8 23
Flax Mill Rd. 10 — K13 38
Flat Pl. 9 — K11 30
Flaxton Pl. 11 — H13 37
Flaxton St. 11 — H13 37
Flaxton Ter. 11 — H13 37
Flaxton Vw. 11
Greenmount Ter. — H13 37
Fleece La. 11 — H12 29
Fleece Sq. 11
Fleece La. — H12 29
Floral Av. 7 — J8 22
Florence Av. 9
Ashley Rd. — K10 30
Florence Gro. 9
Ashley Rd. — K10 30
Florence Mt. 9
Ashley Rd. — K10 30
Florence Pl. 9
Ashley Rd. — K10 30
Florence Rd. 12 — F12 28
Florence St. 9 — L10 31
Florence St. 12
Amberley Rd. — F11 28
Folly La. 11 — H13 37
Forber Pl. 15 — N11 32
Forge La. 12 — F11 28
Forge Pl. 10 — K13 38
Forge Row, 12 — C13 35
Forres St. 7
Meanwood Rd. — G8 21
Forster Pl. 12 — E13 36
Forster's Bldgs. 13 — C9 19
Fortshot La. 17 — L2 9
Foster Ter. 13 — C9 19
Foundry App. 9 — L10 31
Foundry Av. 8 and 9 — L9 23
Foundry Dr. 9 — L9 23
Foundry La.
9 and 14 — M9 to O10 23
Foundry Mill Cres.
14 — N9 24
Foundry Mill Dr. 14 — N9 24
Foundry Mill Gdns.
14 — N8 24
Foundry Mill Mt. 14 — N9 24
Foundry Mill St. 14 — N9 24
Foundry Mill Ter. 14 — N9 24
Foundry Mill Vw. 14 — N9 24
Foundry Mill Wk. 14 — N9 24
Foundry Pl. 9 — L9 23
Foundry St. 11 — H12 29
Foundry Wk. 8 — L9 23
Fountain St. 1 — H11 29
Fountain Wk. 14 — O8 24
Fourteenth Av. 12 — F12 28
Fourth Av. 26 — O15 40
Fox and Grapes Yd. 2
Kirkgate — J11 30
Foxcroft Cl. 6 — E8 20
Foxcroft Grn. 6 — F8 20
Foxcroft Mt. 6 — E8 20
Foxcroft Rd. 6 — E8 20
Foxcroft Wk. 6 — E8 20
Foxhill Av. 16 — F6 12
Foxhill Ct. 16 — F6 12
Foxhill Cres. 16 — F6 12
Foxhill Dr. 16 — F6 12
Foxhill Garth, 16 — F6 12
Foxhill Grn. 16 — F6 12
Foxhill Gro. 16 — F6 12
Foxhill Pl. 16 — F6 12
Foxwood Av. 8 — N8 24
Foxwood Cl. 8 — N8 24

Foxwood Gro. 8 — N8 24
Foxwood Rise, 8 — N8 24
Foxwood Wk. 8 — N8 24
Francis Gro. 11
Lodge La. — H13 37
Francis St. 7 — J9 22
Frankland Pl. 7 — K9 22
Frankland Ter. 7 — K9 22
Fraser Av. 18 — A7 18
Fraser Mt.9 — K10 30
Fraser Rd. 9 — K10 30
Fraser St. 9 — K10 30
Fraser Ter. 9 — K10 30
Frederick Av. 9 — L12 31
Freemantle Pl. 15 — N11 32
Freemont St. 13 — B10 26
Freestone St. 11
Northcote Rd. — H13 37
Front Row, 11 — H12 29
Front St. 11 — H12 29
Fulham Pl. 11 — H13 37
Fulham St. 11 — H13 37
Fulham Ter. 11 — H13 37
Fulneck Ter. 28 — A13 34
Furness Row, 12
Lawns La. — K13 38

Gainsborough Av. 16 — E4 12
Gainsborough Dr. 16 — E4 12
Gaitskell Ct. 11 — H12 29
Gaitskell Gra. 11 — H12 29
Gaitskell Wk. 11 — H12 29
Galway Gro. 11
Hunslet Hall Rd. — H13 37
Galway St. 11 — H13 37
Gamble Hill, 13 — C11 27
Gamble Hill Chase,
13 — C11 27
Gamble Hill Cl. 13 — C11 27
Gamble Hill Cres. 13 — C11 27
Gamble Hill Croft, 13
Gamble Hill View — C11 27
Gamble Hill Fold, 13
Gamble Hill Lawn — C11 27
Gamble Hill Dr. 13 — C11 27
Gamble Hill Gra. 13
Gamble Hill View — C11 27
Gamble Hill Grn. 13 — C11 27
Gamble Hill Lawn,
13 — C11 27
Gamble Hill Pl. 13 — C11 27
Gamble Hill Rise, 13 — C11 27
Gamble Hill Rd. 13 — C11 27
Gamble Hill Vale, 13 — C11 27
Gamble Hill Vw. 13 — C11 27
Gamble La. 12 — C12 27
Ganners Cl. 13 — C9 19
Ganners Garth. 13 — C9 19
Ganners Grn. 13 — C9 19
Ganners Hill, 13 — C9 19
Ganners La. 13 — C9 19
Ganners Mt. 13 — C9 19
Ganners Rise, 13 — C9 19
Ganners Rd. 13 — C9 19
Ganners Wk. 13 — C9 19
Ganners Way, 13 — C9 19
Ganton Mt. 6 — H9 21
Ganton Pl. 6 — H9 21
Ganton Vw. 6 — H9 21
Garden Av. 10
Riley St. — J13 38
Garden Mt. 10
Riley St. — J13 38
Garden Pl. 10
Riley St. — J13 38
Garden St. 9 — J11 30
Garden Ter. 10
Riley St. — J13 38
Garfield Av. 12 — F11 28
Garfield Gro. 12
Armley Rd. — F11 28
Garfield Pl. 13
Harley Rd. — B11 26
Garfield St. 13
Harley Rd. — B11 26
Garfield Ter. 13
Swinnow Rd. — A11 26

Garforth St. 11
Domestic St. — G12 29
Gargrave App. 9 — K11 30
Gargrave Ct. 9 — K11 30
Gargrave Pl. 9 — K10 30
Garland Dr. 15 — P11 33
Garland Fold, 9
Marsh La. — J11 30
Garlick's Yd. 11
Dewsbury Rd. — G15 37
Garmont Rd. 7 — J8 22
Garnet Av. 11
Garnet Pl. — J13 38
Garnet Cres. 11
Burton Rd. — J13 38
Garnet Gro. 11 — J13 38
Garnet Pl. 11 — J13 38
Garnet Rd. 11 — J14 38
Garnet Ter. 11 — J13 38
Garnet Vw. 11 — J13 38
Garr's Ter. 11
Derby Cres. — J12 30
Garth Av. 17 — H6 13
Garth Dr. 17 — H6 13
Garth Rd. 17 — H6 13
Garth, The, 9 — K11 30
Garth Wk. 17 — H6 13
Garton Av. 9 — L11 31
Garton Gro. 9 — L11 31
Garton Rd. 9 — L11 31
Garton Ter. 9 — L11 31
Garton Vw. 9 — L11 31
Gascoigne St. 1 — J11 30
Gateland Dr. 17 — N5 16
Gateland La. 17 — N5 16
Gateside Vw. 15 — N12 32
Gathorne Av. 8
Gathorne Ter. — K9 22
Gathorne Mt. 8 — K9 22
Gathorne St. 8 — K9 22
Gathorne Ter. 8 — K9 22
Gayle's Pl. 11
Greenmount St. — H13 37
Gelderd La. 12 — F13 36
Gelderd Rd. 12 — G12 29
Gelderd Rd. 27
and 12 — C16 to G12 35
Gelder Rd. 12 — E11 28
Genoa Rd. 12 — F11 28
George St. 2 — J11 30
Geranium Pl. 11
Cambrian St. — H13 37
Geranium St. 11
Cambrian St. — H13 37
Geranium Ter. 11
Cambrian St. — H13 37
Ghyll Rd. 6 — E8 20
Gilbert Cl. 5 — E9 20
Gilbert Mt. 5 — E9 20
Gildersome La. 27 — B15 34
Gill Ter. 10
Belinda St. — K13 38
Gill's Sq. 12
Prince St. — G11 29
Gilpin Pl. 12
Lodge Vw. — F12 28
Gilpin St. 12
Lodge Vw. — F12 28
Gilpin Ter. 12
Lodge Vw. — F12 28
Gilpin Vw. 12 — F12 28
Gipsy La. 11 — H15 37
Gipsy La. 26 — O16 40
Gipton App. 9 — M10 31
Gipton Av. 8 — K9 22
Gipton Sq. 9 — M10 31
Gipton St. 8 — K9 22
Gipton Ter. 8 — K9 22
Gipton Vw. 8 — K9 22
Gipton Wood Av. 8 — L8 23
Gipton Wood Cres. 8 — L8 23
Gipton Wood Gro. 8 — L8 23
Gipton Wood Pl. 8 — L8 23
Gipton Wood Rd. 8 — L8 23
Gipton Wood S. 8 — L9 23
Gisburn St. 10 — K13 38
Gladstone St.
Malvern St. — H13 37

Gladstone Ter. 28   A10 26
Gladstone Vw. 11
Malvern St.   H13 37
Gladstone Vils. 17   N5 16
Glasshouse St. 10   J13 38
Glasshouse Vw. 10   H17 42
Glebe Av. 5   E9 20
Glebelands Dr. 6   F8 20
Glebe Pl. 5   E9 20
Glebe St. 28   A12 26
Glebe Ter. 16   F7 20
Gledhow Av. 8   K7 22
Gledhow Ct. 8   K7 22
Gledhow Grange
Vw. 8   K7 22
Gledhow La.
7 and 8   J7 to L8 22
Gledhow Mt. 8   K10 30
Gledhow Park Av. 7   K8 22
Gledhow Park Cres. 7 K8 22
Gledhow Park Dr. 7   J8 22
Gledhow Park Gro. 7 K8 22
Gledhow Park Rd. 7   K8 22
Gledhow Park Vw. 7   K8 22
Gledhow Pl. 8   K10 30
Gledhow Rise, 8   L8 23
Gledhow Rd. 8   K10 30
Gledhow St. 12   F11 28
Gledhow Ter. 8   K10 30
Gledhow Towers, 8   K7 22
Gledhow Valley Rd.
17 and 18   J7 22
Gledhow Wood Av. 8 K7 22
Gledhow Wood
Gro. 8   K7 22
Gledhow Wood Rd. 8 K7 22
Glencoe Av. 9
Glencoe Rd.   K12 30
Glencoe Gro. 9
Glencoe Rd.   K12 30
Glencoe Rd. 9   K12 30
Glencoe St. 9
Glencoe Rd.   K12 30
Glencoe Ter. 9
Glencoe Rd.   K12 30
Glencoe Vw. 9   K12 30
Glenlea Gdns. 13   B9 18
Glen Rd. 16   F7 20
Glensdale Gro. 9   K11 30
Glensdale Mt. 9   K11 30
Glensdale Rd. 9   K11 30
Glensdale St. 9   K11 30
Glensdale Ter. 9   K11 30
Glenthorpe Av. 9   L11 31
Glenthorpe Cres. 9 L11 31
Glenthorpe Ter. 9   L11 31
Globe Rd. 11   H11 29
Glossop Gro. 6   H9 21
Glossop Vw.
Glossop Mt. 6   H9 21
Glossop Vw.
Glossop St. 6   H9 21
Glossop Ter. 6   H9 21
Glossop Vw. 6   H9 21
Gloucester Av. 12   F11 28
Gloucester Gro. 12
Gloucester Rd.   F11 28
Gloucester Pl. 12
Armley Rd.   F11 28
Gloucester Rd. 12   F11 28
Gloucester St. 10   J12 30
Gloucester Ter. 12   G11 29
Gloucester Vw. 12
Gloucester Rd.   F11 28
Golcar Pl. 6
Rider Rd.   H9 21
Golcar St. 6   H9 21
Golden Ter. 12   E13 36
Gold St. 10   J12 30
Goodman St. 10   K12 30
Goodman Ter. 10   J12 30
Goodrick La. 17   H3 7
Goodwin Rd. 12   F12 28
Gordon Dr. 6   G8 21
Gordon Pl. 6
Gordon Ter.   G8 21
Gordon Rd. 10   J13 38
Gordon Ter. 6   G8 21

Gordon Vw. 6
Meanwood Rd.   G8 21
Gotts Park Av. 12   D10 27
Gotts Park Cres. 12 D10 27
Gotts Park Vw. 12   D10 27
Government St. 1
Grace St.   H11 29
Gower St. 2   J11 30
Grace Pl. 12
Thornhill Rd.   E12 28
Grace St. 1   H11 29
Grace Ter. 12   E12 28
Grafton St. 7   J10 30
Graham Av. 4   F9 20
Graham Gro. 4   F9 20
Graham St. 4   F9 20
Graham Ter. 4   F9 20
Graham Vw. 4   F9 20
Granby Av. 6
Granby Ter.   F8 20
Granby Gro. 6   F9 20
Granby Mt. 6
Granby Ter.   F8 20
Granby Pl. 6   F9 20
Granby Rd. 6   F9 20
Granby St. 6
Granby Ter.   F8 20
Granby Ter. 6   F8 20
Granby Vw. 6   F9 20
Grand Arcade, 1
Vicar La.   J11 30
Grand St. 9   K12 30
Grange Av. 7   J9 22
Grange Ct. 6   G8 21
Grange Cres. 7   K9 22
Grange Dr. 18   A7 18
Grangefield Ho.
28   A10 26
Grangefield Rd.
28   A10 26
Grange Park Av. 8   M8 23
Grange Park Cl. 8   N8 24
Grange Park Cres. 8 M8 23
Grange Park Gro. 8   M8 23
Grange Park Pl. 8   M8 23
Grange Park Rise, 8   M8 23
Grange Park Rd. 8   M8 23
Grange Park Ter. 8   M8 23
Grange Park Wk. 8   M8 23
Grange Rd. The, 16   E6 12
Grange Ter. 7   J9 22
Grange Ter. 28   A11 26
Grange Vw. 7   J9 22
Grange Vw. 28   A11 26
Grange Vw. Gdns. 17 N6 16
Grange Villas, The, 16
Otley La.   E6 12
Grange Yd. The, 16   E6 12
Granhamthorpe, 13   C10 27
Granite St. 12
Armley Rd.   F11 28
Granny Av. 27   F15 36
Grantham Towers, 9 K10 30
Granton Rd. 7   J8 22
Granville Rd. 9   K10 30
Grape St. 10   J12 30
Grasmere Av. 12   F11 28
Grasmere Mt. 12
Grasmere Rd.   F12 28
Grasmere Pl. 12   F11 28
Grasmere Rd. 12   F12 28
Grassfield Ter. 18   B7 18
Graveley Ct. 6
Woodhouse St.   H9 21
Graveley Sq. 15
Chapel St.   O11 32
Graveleythorpe Rd.
15   O10 32
Gray Ct. 15   P11 33
Grayrigg Cl. 15   N11 32
Grayrigg Ct. 15   M11 31
Grayrigg Fold, 15   M11 31
Grayrigg Lawn, 15   M11 31
Grayson Cres. 4   E9 20
Grayson Heights, 4   E9 20
Great George St.
2 and 1   H11 29
Great Wilson St. 11 H12 29

Greaves Pl. 10
Orchard St.   K12 30
Greek St. 1   H11 29
Green Bank 11,
Vicar St.   J13 38
Greenbanks Av. 18   C6 11
Greenbanks Cl. 18   C6 11
Greenbanks Dr. 18   B6 10
Green Cl. 6   G7 21
Green Cres. 6   G7 21
Greenfield Av. 27   B15 34
Greenfield Pl. 27   B15 34
Greenfield Rd. 9   K11 30
Green Gdns. 17   H5 13
Greenhead Rd. 16   E7 20
Green Hill Av. 13   D10 27
Green Hill Cl. 12   D10 27
Greenhill Cres. 12   E12 28
Green Hill Dr. 13   D10 27
Greenhill Gdns. 12   E12 28
Green Hill La. 12   E12 28
Green Hill Mt. 13   D10 27
Green Hill Rd. 13   D10 27
Green Hill Vw. 13   D10 27
Green Hill Way, 13   D10 27
Greenhow Av. 4   F10 28
Greenhow Cres. 4   F10 28
Greenhow Gro. 4   F10 28
Greenhow Mt. 4   F10 28
Greenhow Pl. 4   F10 28
Greenhow Rd. 4   F10 28
Greenhow St. 4   F10 28
Greenhow Ter. 4   F10 28
Greenhow Vw. 4
Greenhow St.   F10 28
Greenhow Wk. 4   F10 28
Green La. 11   G14 37
Green La. 12   C12 27
Green La. 12   G12 29
Green La. 14   O7 24
Green La. 15   O10 32
Green La. 16   C5 11
Green La. 18   B7 18
Greenmount Pl. 11   H13 37
Greenmount St. 11   H13 37
Greenmount Ter. 11 H13 37
Greenmount Ter. 11
Marshall St.   H12 29
Greenock Pl. 12   E11 28
Greenock Rd. 12   E11 28
Greenock St. 12
Paisley Rd.   E11 28
Greenock Ter. 12
Paisley Rd.   E11 28
Green Rd. 6   G7 21
Green Rd, W. 6   G7 21
Green Row, 6   G7 21
Greenside Av. 12
Lower Wortley Rd.   E12 28
Greenside Cl. 12   F12 28
Greenside Dr. 12   F12 28
Greenside Rd. 12   F12 28
Greenside Ter. 12   E12 28
Greenside Wk. 12   E12 28
Green, The, 5   D7 19
Green, The, 17   J6 14
Green, The, 27   C15 35
Green, The, 18   B7 18
Greenthorpe Hill,
13   D11 27
Greenthorpe Mt. 13 D11 27
Greenthorpe St. 13   D11 27
Greenthorpe Rd. 13 D11 27
Greenthorpe Wk. 13 D11 27
Green Top, 12   E12 28
Green Vw. 6   G7 21
Greenville Av. 12   E12 28
Greenville Gdns. 12 E12 28
Green Way, 15   P10 33
Green Way Cl. 15   P10 33
Greenwood Mt. 6   G7 21
Greenwood Row,
28   A11 26
Greenwood St. 10   K13 38
Greyhound St. 9
York Rd.   K11 30
Greysheils Av. 6   F9 20
Greystone Mt. 15   N11 32

57

Grimthorpe Pl. 6	F8	20
Grimthorpe St. 6	F8	20
Grimthorpe Ter. 6	F8	20
Grisdale's Yd. 11		
Dewsbury Rd.	G15	37
Grosmont Rd. 13	C10	27
Grosmont Ter. 13		
Grosmont Rd.	C10	27
Grosvenor Av. 7	J10	30
Grosvenor Cres. 7		
Grosvenor Ter.	G9	21
Grosvenor Hill, 7	H9	21
Grosvenor Mt. 6	G9	21
Grosvenor Pk. 7	J7	22
Grosvenor Pl. 7	H10	29
Grosvenor Pl. 10		
Church St.	J13	38
Grosvenor Rd. 6	G9	21
Grosvenor St. 7	H10	29
Grosvenor Ter. 6	G9	21
Grosvenor Ter. 7	H10	29
Grosvenor Vw. 7		
Grosvenor Ter.	H10	29
Grove Av. 28	A11	26
Grove Av. 6	G8	21
Grove Farm Cl. 16	D5	11
Grove Farm Cres. 16	D5	11
Grove Farm Croft, 16	D5	11
Grove Farm Dr. 16	D5	11
Grove Gdns. 6	G8	21
Grovehall Av. 11	G14	37
Grovehall Dr. 11	G14	37
Grovehall Par. 11	G14	37
Grovehall Rd. 11	G14	37
Grove House La. 2	J10	30
Grove La. 6	F8	20
Grove Pl. 2		
Claypit La.	J10	30
Grove Rise, 13	G4	13
Grove Rd. 6	G8	21
Grove Rd. 10	K13	38
Grove Rd. 15	O11	32
Grove Rd. 18	B7	18
Grove Rd. 28	A11	26
Grove St. 1	H11	29
Grove St. 28	A10	26
Grove Ter. 2		
Camp Rd.	J10	30
Grove Ter. 28	A11	26
Grove, The, 17	G4	13
Grove, The, 18	B7	18
Grunberg, Gro. 6		
Bennett Rd.	F8	20
Grunberg Pl. 6		
North La.	F9	20
Grunberg Rd. 6		
Bennett Rd.	F8	20
Grunberg St. 6		
North La.	F9	20
Haddon Pl. 4	F10	28
Haddon Rd. 4	F10	28
Haigh Av. 26	M15	39
Haigh Gdns. 26	M15	39
Haigh Park Rd. 10	L14	39
Haigh Park Vw. 10		
Haigh Park Rd.	L14	39
Haigh Rd. 26	N16	40
Haigh Ter. 26	M15	39
Haigh Vw. 26	M15	39
Haigh Wood, 16	C5	11
Haigh Wood Grn. 16	C5	11
Haigh Wood Rd. 16	C5	11
Halcyon Hill, 7	J7	22
Hales Rd. 12	E12	28
Half Mile, 13	A10	26
Half Mile Cl. 28	A10	26
Half Mile Gdns. 13	A9	18
Half Mile Grn. 28	A10	26
Half Mile La.		
13 and 28	A9	18
Hall Dr. 16	C1	5
Hall Gro. 6	G10	29
Halliday Av. 12	E11	28
Halliday Dr. 12	E11	28
Halliday Gro. 12	E11	28
Halliday Mt. 12	E11	28

Halliday Pl. 12	E11	28
Halliday St. 28	A11	26
Hall La. 7	J9	22
Hall La. 12	C12	27
Hall La. 12	F11	28
Hall La. 18	A7	18
Hall Park Av. 18	B6	10
Hall Park Cl. 18	B6	10
Hall Park Garth, 18	B6	10
Hall Park Mt. 18	B6	10
Hall Park Rise, 18	B6	10
Hall Pl. 9	K11	30
Halton Moor Av. 9	M12	31
Halton Moor Rd.		
9 and 15	L12 to N12	31
Halton Rd. 15	N11	32
Hamerton St. 28	A11	26
Hamilton Av. 7	K9	22
Hamilton Pl. 7	K9	22
Hamilton Ter. 7	K9	22
Hamilton Vw. 7	K9	22
Hampshire Ter. 7		
Claypit La.	J10	30
Hampton Pl. 9	K11	30
Hampton Ter. 9	K11	30
Hanover Av. 3	G10	29
Hanover La. 3	H11	29
Hanover Mt. 3		
Hanover Sq.	H10	29
Hanover Sq. 3	H10	29
Hanover St. 3	H11	29
Hanover St. 13		
Upper Town St.	C9	19
Hanover Vw. 3		
Hanover Av.	G10	29
Hansby Av. 14	O8	24
Hansby Bank, 14	O8	24
Hansby Cl. 14	O9	24
Hansby Dr. 14	O8	24
Hansby Gdns. 14	O9	24
Hansby Garth, 14	O8	24
Hansby Gra. 14	O8	24
Hansby Pl. 14	O8	24
Harcourt Pl. 1	H11	29
Harcourt St. 1		
Grace St.	H11	29
Hardistry's Yd. 10		
Waterloo Rd.	K13	38
Hardrow Gro. 12		
Highfield Gro.	F12	28
Hardrow Rd. 12	F12	28
Hardrow Ter. 12		
Highfield Gro.	F12	28
Hardwick St. 10	K14	38
Hardwick's Yd. 1		
Briggate, 1	J11	30
Hardy Gro. 11		
Wickham St.	H13	37
Hardy St. 11	H13	37
Hardy Ter. 11		
Lodge La.	H13	37
Harefield E, 15	N11	32
Harefield W. 15	N11	32
Harehills Av.		
7 and 8	K9	22
Harehills La. 7, 8		
and 9	K8 to M10	22
Harehills Park Av. 9	L10	31
Harehills Park Rd. 9	L10	31
Harehills Park Ter. 9	L10	31
Harehills Park Vw. 9	L10	31
Harehills Pl. 8	K9	22
Harehills Rd. 8	K9	22
Harehills Ter. 8	K9	22
Hare La. 28	A12	26
Hare Park Mt. 12	C11	27
Hares Av. 8	K9	22
Hares Mt. 8	K9	22
Hares Rd. 8	K9	22
Hares Ter. 8	K9	22
Hares Vw. 8	K9	22
Harewood St. 2	J11	30
Harlech Av. 11	H14	37
Harlech Cres. 11	H14	37
Harlech Gro. 11	H14	37
Harlech Mt. 11		
Harlech Rd.	H14	37
Harlech Rd. 11	H14	37

Harlech St. 11	H14	37
Harlech Ter. 11	H14	37
Harley Ct. 13	B11	26
Harley Dr. 13	B11	26
Harley Gdns. 28	B11	26
Harley Grn. 13	B11	26
Harley Rise, 13	B11	26
Harley Rd. 13	B11	26
Harley St. 13	B11	26
Harley Ter. 13	B11	26
Harley Vw. 13	B11	26
Harley Wk. 13	B11	26
Harold Av. 6	G10	29
Harold Gro. 6	G10	29
Harold Mt. 6	G10	29
Harold Pl. 6	G10	29
Harold Rd. 6	G10	29
Harold St. 6	G10	29
Harold Ter. 6	G10	29
Harold Vw. 6	G10	29
Harold Wk. 6	G10	29
Harper Pl. 2		
Harper St.	J11	30
Harper St. 2	J11	30
Harriet St. 7	J9	22
Harrison Cres. 9	M10	31
Harrison's Av. 28	A10	26
Harrison St. 1	J11	30
Harrogate Rd.		
7 and 17	J8 to L1	22
Harrogate Rd. 19	A3	4
Harrowby Cres. 16	E7	20
Harrowby Rd. 16	E7	20
Harrow St. 12	E11	28
Harry Cres. 9	K11	30
Harthill Av. 27	C15	35
Harthill La. 27	C15	35
Hartington Rd. 12	F12	28
Hartington St. 12	F12	28
Hartley Av. 6	H9	21
Hartley Cres. 6	H9	21
Hartley Gro. 6	H9	21
Hartley St. 27	E15	36
Hartley Ter. 11	G12	29
Harton Ter. 11		
Cemetery Rd.	H13	37
Hartwell Pl. 6	G10	29
Hartwell Rd. 6	G10	29
Harwill Av. 27	F15	36
Haslewood Cl. 9	K11	30
Haslewood Ct. 9	K11	30
Haslewood Dene, 9	K11	30
Haslewood Dr. 9	K11	30
Haslewood Grn. 9	K11	30
Haslewood Lawn, 9	K11	30
Haslewood Ms. 9	K11	30
Haslewood Pl. 9	K11	30
Haslewood Sq. 9	K11	30
Hatfield St. 9		
Shannon St.	K11	30
Haven, The, 15	P11	33
Havercroft, 12		
Cross La.	E11	28
Hawes Mt. 6	H9	21
Hawes Pl. 6	H9	21
Hawes St. 6	H9	21
Hawes Ter. 6	H9	21
Hawkhill Av. 15	O10	32
Hawkhill Dr. 15	O10	32
Hawkhill Gdns. 15	O10	32
Hawkhurst Rd. 12	F12	28
Hawkins Pl. 7	J10	30
Hawkins St. 7	J10	30
Hawkshead Cres. 14	N9	24
Hawks Nest Gdns.		
E. 17	J5	14
Hawks Nest Gdns.		
S. 17	J5	14
Hawks Nest Gdns.		
W. 17	J5	14
Hawks Nest Rise, 17	J5	14
Hawkswood Av. 5	D7	19
Hawkswood Cres. 5	D7	19
Hawkswood Gro. 5	D8	19
Hawkswood Mt. 5	D7	19
Hawkswood Pl. 5	D8	19
Hawkswood St. 5	D8	19
Hawkswood Ter. 5	D8	19

Hawkswood Vw. 5   D7   19
Hawksworth Gro. 5   C8   19
Hawksworth Rd.
   18   C7   19
Hawthorn Av. 12   G11   29
Hawthorn Gro. 12   G11   29
Hawthorn Mt. 7   J7   22
Hawthorn Pl. 12
   Hall La.   F11   28
Hawthorn Rd. 7   J7   22
Hawthorn St. 11
   Cambrian St.   H13   37
Hawthorn Ter. 12
   Hall La.   F11   28
Hawthorn Vale, 7
   Hawthorn Rd.   J7   22
Hawthorn Vw. 7   J7   22
Hawthorn Wk. 7   J7   22
Haydn's Ter. 28
   Arthur St.   A10   26
Hayleigh St. 13
   Warrel's Rd.   C9   19
Hayleigh Ter. 13
   Warrel's Rd.   C9   19
Haymount Pl. 9
   Hill St.   K10   30
Haymount Row, 9
   Hill St.   K10   30
Haymount Sq. 9
   Hill St.   K10   30
Haymount St. 9
   Hill St.   K10   30
Hazelhead St. 10
   Clarence Rd.   J12   30
Headingley Av. 6   F8   20
Headingley Cres. 6   F9   20
Headingley La. 6   G9   21
Headingley Mt. 6   F8   20
Headingley Ter.
   Grosvenor Rd.   G9   21
Headingley Vw. 6   F9   20
Headrow, The, 1   H11   29
Heath Cres. 11   G13   37
Heathcroft Bank, 11   G14   37
Heathcroft Cres. 11   G14   37
Heathcroft Dr. 11   G14   37
Heathcroft Lawn, 11   G14   37
Heathcroft Rise, 11   G14   37
Heathcroft Vale, 11   G14   37
Heather Gdns. 13   D11   27
Heather Gro. 13   C11   27
Heathfield, 16   E5   12
Heathfield Ter. 6   F8   20
Heath Gro. 11   G13   37
Heath Mt. 11   G13   37
Heath Pl. 11   G13   37
Heath Rise, 11   G13   37
Heath Rd. 11   G13   37
Heaton Av. 12   F12   28
Heaton's Ct. 1
   Briggate   J11   30
Hebden App. 14   O8   24
Hebden Grn. 14   P8   25
Hebden Pl. 14   O8   24
Heddon Pl. 6
   Brookfield Rd.   G8   21
Heddon St. 6
   Brookfield Rd.   G8   21
Hedley Chase, 12   G11   29
Hedley Gdns. 12   G11   29
Hedley Grn. 12   G11   29
Heights Bank, 12   D11   27
Heights Cl. 12   D11   27
Heights Dr. 12   D11   27
Heights Garth, 12   D11   27
Heights Grn. 12   D11   27
Heights La. 12   D11   27
Heights Par. 12   D11   27
Heights, The, W. 12   D11   27
Heights Wk. 12   D11   27
Heights Way, 12   D11   27
Helmscott Garth, 15   Q9   25
Helmsley Dr. 16   E7   20
Helmsley Rd. 16   E7   20
Helston Cl. 10   H16   42
Helston Croft, 10   H16   42
Helston Garth, 10   H16   42
Helston Grn. 10   H16   42

Helston Nook, 10   H16   42
Helston Pl. 10   H16   42
Helston Rd. 10
       G16 to H16   42
Helston Sq. 10   G16   42
Helston St. 10   G16   42
Helston Wk. 10   H16   42
Helwood La. 14   P3   42
Hemingway Cl. 10   K13   38
Hemingway Garth,10   K13   38
Hemingway Gro. 10   K13   38
Henbury St. 7   J10   30
Henconner Av. 7   J8   22
Henconner Cres. 7   J8   22
Henconner Dr. 7   J8   22
Henconner Gdns. 7   J8   22
Henconner Gro. 7   J8   22
Henconner La. 13   D11   27
Henconner La. 7   J8   22
Henconner Rd. 7   J8   22
Henley Cres. 13   C10   27
Henley Gro. 13
   Henley Rd.   C10   27
Henley Pl. 13
   Henley Rd.   C10   27
Henley Rd. 13   C10   27
Henley St. 13
   Henley Rd.   C10   27
Henley Vw. 13   C10   27
Henry Av. 12   E12   28
Henry St. 1   H11   29
Hepworth Av. 27   E15   36
Herbalist St. 12
   Tong Rd.   E11   28
Herbert Gro. 7
   Oatland Rd.   J10   30
Herbert Pl. 7   J10   30
Herbert St. 7   J9   22
Herbert Ter. 7   J10   30
Hermon Rd. 15   O10   32
Hermon St. 15   O10   32
Hertford St. 10
   Waterloo Rd.   K13   38
Hesketh Av. 5   E9   20
Hesketh Mt. 5   E8   20
Hesketh Pl. 5   E8   20
Hesketh Rd. 5   E8   20
Hesketh Ter. 5   E9   20
Hessle Av. 6   G9   21
Hessle Mt. 6   G9   21
Hessle Pl. 6   G9   21
Hessle St. 6
   Walmlsey Rd.   G9   21
Hessle Ter. 6   G9   21
Hessle Vw. 6   G9   21
Hessle Wk. 6   G9   21
Hetton Rd. 8   L8   23
Hetton Rd. S. 8   L9   23
Heywood Lawn. 10   J13   38
Heywood Pl. 10
   Hulland St.   J13   38
Heywood St. 10
   Hulland St.   J13   38
High Ash Av. 17   K4   14
High Ash Cres. 17   K4   14
High Ash Dr. 17   K4   14
High Ash Mt. 17   K4   14
Highbury Gro. 13
   Harley Rd.   B11   26
Highbury La. 6
   Monk Bridge Rd.   G8   21
Highbury Mt. 6   G8   21
Highbury Pl. 6   G8   21
Highbury Pl. 13   B11   26
Highbury Rd. 6   G8   21
Highbury St. 6   G8   21
Highbury St. 13
   Harley Rd.   B11   26
Highbury Ter. 6   G8   21
Highbury Ter. 13
   Harley Rd.   B11   26
High Cliffe, 4
   St. Michael's La.   F9   20
High Court La. 2   J11   30
Highfield, 27   D15   35
Highfield Av. 12   F12   28
Highfield Cl. 12   F12   28
Highfield Cres. 12   F12   28

Highfield Gdns. 12   F12   28
Higefield Gro. 12   F12   28
Highfield La. 26   P15   41
Highfield Mt. 11
   Highfield St.   H13   37
Highfield Pl. 12
   Bruce St.   G11   29
Highfield Rd. 13   C10   27
Highfield St. 13
   Lower Town St.   C10   27
Highfield Ter. 11
   Highfield St.   H13   37
Highfield Ter. 13
   Lower Town St.   C10   27
Highgate St. 10
   Low Rd.   K13   38
High Markland St. 9
   Ellerby Rd.   K11   30
Highmoor Av. 17   K5   14
Highmoor Cres. 17   K5   14
Highmoor Dr. 17   K5   14
Highmoor Gro. 17   K5   14
High Ridge, 26   N15   40
High Ridge Av.
   26   M15   39
High Ridge Pk.
   26   M15 to N15   39
High Ridge Way, 16   D2   5
High St. 15   O11   32
High St. 13   A9   18
Highthorne Dr. 17   K5   14
Highthorne Gro. 12
   Highthorne St.   E11   28
Highthorne Mt. 17   K5   14
Highthorne St. 12   E11   28
Highthorne Ter. 17   K5   14
Highthorne Vw. 12
   Highthorne St.   E11   28
Highways, 14   N10   32
Highwood Av. 17   J5   14
Highwood Cres. 17   J5   14
Hillam St. 11
   Domestic St.   G12   29
Hillcourt Av. 13   C9   19
Hillcourt Dr. 13   C9   19
Hillcourt Gro. 13   C9   19
Hillcrest Av. 7   K9   22
Hillcrest Pl. 7   K9   22
Hillcrest Rise, 16   D5   11
Hillcrest Vw. 7   K9   22
Hill End, 12   D11   27
Hill End Cres. 12   D11   27
Hill End Rd. 12   D11   27
Hillidge Rd. 10   J13   38
Hillidge Sq. 10   J13   38
Hillllngdon Way, 17   I14   13
Hill Rise Av. 13   C9   19
Hill Rise Gro. 13   C9   19
Hill Side Mt. 28
   Arthur St.   A10   26
Hillside Rd. 7   J7   22
Hillside St. 9
   Hill St.   K10   30
Hill Side Vw. 28
   Arthur St.   A10   26
Hill St. 11   H13   37
Hill St. 9   K10   30
Hill's Yd. 11   H12   29
Hillthorpe Rd. 28   A12   26
Hillthorpe Sq. 28   A12   26
Hillthorpe St. 28   A12   26
Hillthorpe Ter. 28   A12   26
Hill Top, 12
   Stanningley Rd.   D10   27
Hill Top Av. 8   K9   22
Hill Top Cl. 12   D11   27
Hill Top Mt. 8   K9   22
Hill Top Rd. 6   G10   29
Hill Top Rd. 12   D11   27
Hill View Av. 7
   Norfolk Gdns.   J7   22
Hill View Mt. 7
   Pasture La.   J7   22
Hill View Pl. 7
   Pasture La.   J7   22
Hill View Ter. 7
   Pasture La.   J7   22

Hilton Gro. 8	K9	22
Hilton Pl. 8	K9	22
Hilton Rd. 8	K9	22
Hird St. 11		
Tempest Rd.	H13	37
Hirst's Yd. 1		
Briggate	J11	30
Hobberley La. 17	N5	16
Hodgson Av. 17	L5	15
Hodgson Cres. 17	L5	15
Holbeck La. 11	G12	29
Holbeck Moor Rd. 11	H12	29
Holborn St. 6	H9	21
Holborn Ter. 6	H9	21
Holborn Towers, 6	H9	21
Holderness Ter. 6		
Kings Rd.	G10	29
Holdforth Chase, 12	G11	29
Holdforth Gdns. 12	G11	29
Holdforth Grn. 12	G11	29
Holdforth Pl. 12	G11	29
Holdforth Sq. 9		
Cotton St.	J11	30
Holdsworth's Fold, 9		
Richmond St.	K11	30
Hollin Cres. 16	F7	20
Hollin Dr. 16	F7	20
Hollin Gdns. 16	F7	20
Hollin Hall Av. 8	M8	23
Hollin Hill Dr. 8	M8	23
Hollin La. 16	F7	20
Hollin Mt. 16	F7	20
Hollin Park Av. 8	M8	23
Hollin Park Cres. 8	M8	23
Hollin Park Mt. 8	M8	23
Hollin Park Pl. 8	M8	23
Hollin Park Rd. 8	M8	23
Hollin Park Ter. 8	M8	23
Hollin Park Vw. 8	M8	23
Hollin Pl. 16		
Weetwood La.	F6	12
Hollin Rd. 16	F7	20
Hollin Vw. 16	F7	20
Holly Av. 16	C5	11
Holly Dr. 16	C5	11
Hollyshaw Cres. 15	P11	33
Hollyshaw Gro. 15	P11	33
Hollyshaw La. 15		
	P10 to P11	33
Hollyshaw St. 15	P11	33
Hollyshaw Ter. 15		
Hollyshaw St.	P11	33
Hollyshaw Walk, 15	P10	33
Hollywell La. 12	E11	28
Holmesley Cl. 26	O16	40
Holmesley Field La.		
26	O16	40
Holmesley Garth,		
26	O15	40
Holmesley La. 26	O16	40
Holmes St. 11	J12	30
Holmfield Dr. 8	K6	14
Holmwood Av. 6	G7	21
Holmwood Cl. 6	G7	21
Holmwood Cres. 6	G7	21
Holmwood Dr. 6	G7	21
Holmwood Gro. 6	G7	21
Holmwood Mt. 6	G7	21
Holmwood Vw. 6	G7	21
Holroyd St. 7	J10	30
Holroyd Yd. 11		
Meadow La.	J12	30
Holt Av. 16	F4	12
Holt Cl. 16	F4	12
Holt Gdns. 16	F4	12
Holt Vw. 16	F4	12
Holt La. 16	D4	11
Holywell La. 17	M4	15
Holywell Vw. 17		
Holywell La.	M4	15
Home Lea, 26	M15	39
Hope Av. 12	F11	28
Hope Cres. 12		
Stanningley Rd.	D10	27
Hopeful View, 10	J16	43
Hope Gro. 12		
Stanningley Rd.	D10	27
Hope Gro. 12	F11	28

Hope Mt. 12	F11	28
Hope Pl. 2	J11	30
Hope Pl. 12	F11	28
Hope Vw. 12		
Stanningley Rd.	D10	27
Hopewell Pl. 6		
Harold St.	G10	29
Hopewell Ter. 10		
Glasshouse St.	J13	38
Hopewell Vw. 10	J16	43
Hopwood Rd. 18	C6	11
Horsforth New Rd.		
13	A8	18
Hospital Lane, 16	D5	11
Hough End Av. 13	C10	27
Hough End Cl. 13	C10	27
Hough End Ct. 13	C10	27
Hough End Cres. 13	C10	27
Hough End Garth, 13	C10	27
Hough End La. 13	C10	27
Hough Cl. 13	B11	26
Hough La. 13	C10	27
Houghley Av. 12	D10	27
Houghley Cl. 13	D10	27
Houghley Cres. 12	D10	27
Houghley Gro. 12	D10	27
Houghley La. 13	D10	27
Houghley Mt. 12	D10	27
Houghley Pl. 12	D10	27
Houghley Rd. 12	D10	27
Houghley Sq. 12	D10	27
Houghley St. 12	D10	27
Houghley Ter. 12	D10	27
Houghley Vw. 12	D10	27
Hough Side Cl. 28	B11	26
Hough Side La. 28	B11	26
Hough Side Rd. 28	A11	26
Houghton Gro. 11		
Houghton Pl.	H12	29
Houghton St. 11	H13	37
Hough Top, 13	B11	26
Hough Tree Rd. 13	C11	27
Hough Tree Ter. 13		
Pudsey Rd.	C11	27
Hougomont St. 11		
Wortley La.	G12	29
Hovingham Av. 8	L9	23
Hovingham Gro. 8	L9	23
Hovingham Mt. 8		
Hovingham Av.	L9	23
Hovingham Ter. 8		
Hovingham Av.	L9	23
Howard Av. 15	N11	32
Howard Ct. 15	N11	32
Howden Gro. 6	G10	29
Howden Pl. 6	G10	29
Howden St. 6	G10	29
Howden Ter. 6	G10	29
Hoxton Gro. 11	G13	37
Hoxton Mt. 11	G13	37
Hoxton Pl. 11	G13	37
Hoxton Ter. 11	G13	37
Hudson Rd. 9	L10	31
Hudson Sq. 5		
Commercial Rd.	E9	20
Hudson St. 9		
Hudson Rd.	L10	31
Hudson Ter. 10		
Beza St.	J13	38
Humane Pl. 10		
Low Rd.	K13	38
Hunger Hills Av. 18	B6	10
Hunger Hills Dr. 18	B6	10
Hunslet Hall Rd. 11	H13	37
Hunslet La. 10	J12	30
Hunslet Rd. 10	J11	30
Huntley Pl. 11		
Chester Pl.	H13	37
Hustler's Row, 6	F7	20
Hutchinson's Sq. 11		
Dewsbury Rd.	G15	37
Huttons Row, 6		
Green Rd.	G7	21
Hyde Park Corner, 6	G9	21
Hyde Park Rd. 6	G10	29
Hyde Park Ter. 6	G9	21
Hyde St. 2	H10	29
Hyde Ter. 2	H10	29

Ida Cres. 10	L14	39
Ida Gro. 10	L14	39
Ida Mt. 10	L14	39
Ida St. 10	L14	39
Ida Ter. 10	L14	39
Ida Vw. 10	L14	39
Infirmary St. 1	H11	29
Ingham St. 10		
Hunslet Rd.	J11	30
Ingham St. 28		
Cavendish Pl.	A10	26
Ingleby Pl. 12		
Ingleby St.	G11	29
Ingleby Ter. 12		
Ingleby St.	G11	29
Ingledew Cres. 8	L6	15
Ingledew Dr. 8	L6	15
Ingleton Dr. 15	N11	32
Ingleton Gro. 11	H13	37
Ingleton St. 11	H13	37
Ingleton Ter. 11	H13	37
Inglewood App. 14	O9	24
Inglewood Dr. 14	O9	24
Inglewood Pl. 14	O9	24
Inglewood Ter. 6		
Delph La.	H9	21
Ingram Av. 15	N11	32
Ingram Cres. 11		
Ingram Rd.	G12	29
Ingram Rd. 11	G12	29
Ingram Row, 11	H12	29
Ingram St. 11	H12	29
Ingram Vw. 11	G12	29
Ings Cres. 9	L11	31
Ings Rd. 9	L11	31
Institution Row, 6		
Raglan Rd.	H9	21
Institution St. 6	H9	21
Intake La. 10	J17	43
Intake La. 14	P5	17
Intake Mt. 10	J17	43
Intake Rd. 28	A11	26
Intake Sq. 10	J17	43
Intake Vw. 13	B9	18
Intake Vw. 10	J17	43
Iron St. 10	K13	38
Ironwood App. 14	O9	24
Ironwood Cres. 14	O9	24
Ironwood Vw. 14	O9	24
Irvin App. 15	N11	32
Island Ter. 11	G12	29
Isle La. 11		
Balm Wk.	G12	29
Iveridge Gro. 10	K13	38
Iveridge Mt. 10		
Iveridge St.	K13	38
Iveridge St. 10	K13	38
Iveridge Ter. 10	K13	38
Iveson App. 16	D6	11
Iveson Cl. 16	D6	11
Iveson Cres. 16	D6	11
Iveson Dr. 16	D6	11
Iveson Gdns. 16	D6	11
Iveson Garth, 16	E6	12
Iveson Grn. 16	D6	11
Iveson Lawn, 16	E6	12
Iveson Rise, 16	E6	12
Iveson Rd. 16	D6	11
Iveson Wk. 16	D6	11
Ivory St. 10	J12	30
Ivy Av. 9	L11	31
Ivy Cres. 9	L11	31
Ivy Gro. 9	L11	31
Ivy Mt. 9	L11	31
Ivy Pl. 13	C9	19
Ivy Rd. 9	L11	31
Ivy St. 9	L11	31
Ivy Vw. 9	L11	31
Jack La. 11 and 10	H12	29
Jackman Dr. 18	C7	19
Jackman Sq. 6		
Institution St.	H9	21
Jackman Yd. 6		
Institution St.	H9	21
Jackson Av. 8	K7	22

61

Lambert's Yd. I
Briggate J11 30
Lambert Ter. 18 B7 18
Lambeth St. 8 K9 22
Lambeth Ter. 8 K9 22
Lambrigg Cres. 14 O9 24
Lambton Gro. 8
Roundhay Rd. J9 22
Lambton Pl. 8
Roundhay Rd. J9 22
Lambton Vw. 8
Roundhay Rd. J9 22
Lanark Dr. 18 B5 10
Lancaster Av. 5 D9 19
Lancaster Gro. 5 D9 19
Land Ct. 11
Water La. H12 29
Landseer Av. 13 D9 19
Landseer Cl. 13 C9 19
Landseer Cres. 13 D9 19
Landseer Dr. 13 C9 19
Landseer Grn. 13 C9 19
Landseer Gro. 13 D9 19
Landseer Mt. 13 D9 19
Landseer Rise, 10 C9 19
Landseer Rd. 13 C9 19
Landseer Ter. 13 D9 19
Landseer Vw. 13 C9 19
Landseer Wk. 13 C9 19
Landseer Way, 13 C9 19
Lands La. I J11 30
Lane End, 11 H12 29
Lane End, 28 A11 26
Lane End Mt. 28 A11 26
Lanes, The, 28 A11 26
Lane, The, 17 G4 13
Lane, The, 9 K11 30
Langbar App. 14 P8 25
Langbar Cl. 14 P8 25
Langbar Gdns. 14 P8 25
Langbar Garth, 14 P8 25
Langbar Grn. 14 P8 25
Langbar Gro. 14 P8 25
Langbar Pl. 14 P8 25
Langbar Rd. 14 P8 25
Langbar Sq. 14 P8 25
Langbar Vw. 14 P8 25
Langdale Av. 6 F9 20
Langdale Gdns. 6 F9 20
Langdale Ter. 6 F9 20
Langley Av. 13 B9 18
Langley Cl. 13 B9 18
Langley Cres. 13 B9 18
Langley Garth, 13 B9 18
Langley Mt. 13 B9 18
Langley Pl. 13 B9 18
Langley Rd. 13 B9 18
Langthorpe Pl. 11
Cemetery Rd. H13 37
Lansdowne Mt. 11
Cemetery Rd. H13 37
Lansdowne St. 12
Amberley Rd. F11 28
Lanshaw Cl. 10 K16 43
Lanshaw Cres. 10 K16 43
Lanshaw Pl. 10 K16 43
Lanshaw Rd. 10 K16 43
Lanshaw Ter. 10 K16 43
Lanshaw Vw. 10 K16 43
Lanshaw Wk. 10 K16 43
I'Anson Pl. 11
Dewsbury Rd. G15 37
Lapish St. 6
Woodhouse St. H9 21
Larchfield Dene, 10 K13 38
Larkhill Cl. 8 K7 22
Larkhill Grn. 8 K6 14
Larkhill Rd. 8 K6 14
Larkhill Vw. 8 K7 22
Larkhill Wk. 8 K6 14
Larkhill Way, 8 K6 14
Lascelles Mt. 8
Lascelles Rd. K9 22
Lascelles Pl. 8
Lascelles Rd. K9 22
Lascelles Rd. 8 K9 22
Lascelles St. 8
Roundhay Rd. J9 22
Lascelles Ter. 8 K9 22

Lascelles Vw. 8
Lascelles Rd. K6 14
Lastingham Rd. 18 A8 18
Latch Gdns. 16 E6 12
Latchmere Av. 16 D7 19
Latchmere Cres. 16 D7 19
Latchmere Cl. 16 E7 20
Latchmere Cross, 16 D7 19
Latchmere Dr. 16 D7 19
Latchmere Grn. 16 D7 19
Latchmere Rd. 16 D7 19
Latchmere Vw. 16 D6 11
Latch Wk. 16 E6 12
Laura St. 12
Sutton St. G12 29
Laurel Bank Cl. 6 F9 20
Laurel Mt. 28 A11 26
Laurel Mt. 7 J8 22
Laurel Pl. 12 F11 28
Laurel Ter. 12 P9 25
Laurel Ter. 12 F11 28
Lavender Wk. 9 K11 30
Lavinia St. 7 J10 30
Lawefield Av. 26 M16 39
Lawn's La. 10 K13 38
Lawns La. 12 C13 35
Lawnswood Gdns. 16 E6 12
Lawrence Av. 8 M9 23
Lawrence Cres. 8 M9 23
Lawrence Gdns. 8 M8 23
Lawrence Rd. 8 M9 23
Lawrence Wk. 8 M9 23
Lawson St. 12
Privilege St. E11 28
Layton Av. 19 A5 10
Layton Dr. 19 A5 10
Layton La. 19 A6 10
Layton Rd. 19 A5 10
Lea Farm Cres. 5 D8 19
Lea Farm Dr. 5 D7 19
Lea Farm Gro. 5 D8 19
Lea Farm Mt. 5 D7 19
Lea Farm Pl. 5 D8 19
Lea Farm Rd. 5 D7 19
Lea Farm Row, 5 D8 19
Lea Farm Wk. 5 D7 19
Leafield Cl. 17 H6 13
Leafield Dr. 17 H6 13
Lea Ter. 17 H6 13
Leah Pl. 12 G12 29
Leasowe Av. 10 K14 38
Leasowe Gro. 10 K14 38
Leasowe Mt. 10 K14 38
Leasowe Rd. 10 K14 38
Leasowe St. 10
Hardwick St. K14 38
Leasowe Ter. 10
Telford Ter. K13 38
Leasowe Vw. 10
Leasowe Rd. K14 38
Leathley Rd. 10 J12 30
Leathley St. 10 J12 30
Ledbury St. 11
Kirkland St. H12 29
Ledsham Dene, 10 J13 38
Ledsham Pl. 10
Bleasby St. J13 38
Ledsham St. 10
Grove Rd. K13 38
Leeds and Bradford Rd.
28 and 13 A10 to D9 26
Leeds Ter. 7 J10 30
Leek Chase, 10 K13 38
Lee Lane, 18 B6 10
Lee Lane E. 18 B6 10
Lee Lane W. 18 A6 10
Leeming Sq. I
Cankerwell La. H10 29
Lees Yd. 11 J12 30
Leicester Av. 2
Leicester Ter. H10 29
Leicester Gro. 7 H10 29
Leicester Mt. 2
Leicester Ter. H10 29
Leicester Pl. 2 and 7 H10 29
Leicestershire St. 10 K13 38
Leicester Ter. 2 H10 29
Leighton La. I H11 29

Lenhurst Av. 12 D9 19
Lent Pl. 7 J9 22
Lennox Rd. 4 F10 28
Leonard Pl. 7
Rayner St. K9 22
Leonard Ter. 7
Armanda St. K9 22
Leopold Rd. 7 K9 22
Leopold Sq. 7
Leopold St. J9 22
Leopold St. 7 J9 22
Leopold Ter. 7 J9 22
Lepton Pl. 27 C15 35
Leslie Ter. 6
Woodhouse St. H9 21
Levens Bank, 15 M12 31
Levens Cl. 15 N12 32
Levens Garth, 15 N12 32
Levens Pl. 15 N12 32
Levita Pl. 15 N11 32
Lewis Av. 12 F12 28
Lewis Pl. 12 F12 28
Leyburn Pl. 11
Cemetery Rd. H13 37
Leyburn St. 11
Cemetery Rd. H13 37
Leyburn Ter. 11 H13 37
Ley La. 12 F11 28
Leysholme Cres. 12 E12 28
Leysholme Dr. 12 E12 28
Leysholme Ter. 12 E12 28
Leysholme Vw. 12 E12 28
Lickless Av. 18 C6 11
Lickless Dr. 18 C6 11
Lickless Gdns. 18 C6 11
Lickless Ter. 18 C6 11
Lidget Hill, 28 A11 26
Lidgett Av. 8 K7 22
Lidgett Ct. 8 K7 22
Lidgett Cres. 8 K7 22
Lidgett Gdns. 8 K7 22
Lidgett Gro. 8 K7 22
Lidgett La. 17 and 18 J6 14
Lidgett Mt. 8 K6 14
Lidgett Park Av. 8 K6 14
Lidgett Park Gdns. 8 K7 22
Lidgett Park Gro. 8 K6 14
Lidgett Park Rd. 8 K6 14
Lidgett Pl. 8 K7 22
Lidgett Row, 8 K6 14
Lidgett Towers, 8 K6 14
Lidgett Vw. 8 K7 22
Lidgett Wk. 8 K7 22
Lilac Gro. 7
Skinner La. J10 30
Lilac Pl. 10
Arthington Av. J14 38
Lilac St. 2 J10 30
Lilian Pl. 4 F10 28
Lilian St. 4 F10 28
Lime Tree Av. 17 K6 14
Limewood App. 14 O7 24
Limewood Rd. 14 O8 24
Lincoln Av. 9 K10 30
Lincoln Green Rd. 9 K10 30
Lincoln Gro. 9
Lincoln Rd. K10 30
Lincoln Mt. 9 K10 30
Lincoln Rd. 9 K10 30
Lincoln Ter. 9
Lincoln Rd. K10 30
Lincoln Towers, 9 K10 30
Lincoln Vw. 9 K10 30
Lincombe Bank, 8 K7 22
Lincombe Dr. 8 K7 22
Lincombe Mt. 8 K7 22
Lincombe Rise, 8 K7 22
Lincroft Cres. 13 C9 19
Linden Gro. 11 J13 38
Linden Mt. 11
Linden Rd. H13 37
Linden Pl. 11
Linden Rd. H13 37
Linden Rd. 11 H13 37
Linden St. 11
Linden Rd. H13 37
Linden Ter. 11 J13 38
Lindsey Ct. 9 K10 30

Lindsey Gdns. 9	K10	30
Lindsey Mt. 9	K10	30
Lindsey Rd. 9	K10	30
Lingfield App. 17	H5	13
Lingfield Bank, 17	H5	13
Lingfield Cl. 17	J5	14
Lingfield Cres. 17	H5	13
Lingfield Dr. 17	J5	14
Lingfield Gdns. 17	H5	13
Lingfield Gate, 17	H5	13
Lingfield Grn. 17	H5	13
Lingfield Gro. 17	J5	14
Lingfield Hill, 17	H5	13
Lingfield Mt. 17	J5	14
Lingfield Par. 17	J5	14
Lingfield Rd. 17	H5	13
Lingfield Vw. 17	H5	13
Lingfield Wk. 17	H5	13
Lingwell Av. 10	J16	42
Lingwell Cres. 10	J16	43
Lingwell Gro. 10	J17	43
Lingwell Rd. 10	J16	43
Lingwell Sq. 10	J16	43
Lingwell St. 10	J16	43
Linton Av. 17	K5	14
Linton Cl. 17	K5	14
Linton Cres. 17	K5	14
Linton Dr. 17	K5	14
Linton Gro. 17	K5	14
Linton Rise, 17	K5	14
Linton Rd. 17	K5	14
Lisbon St. 1	H11	29
Lister Hill, 18	C6	11
Little La. 27	F15	36
Littlemoor Bottom, 28	A12	26
Littlemoor Cres. 28	A12	26
Littlemoor Gdns. 28	A12	26
Littlemoor La. 14	Q5	17
Littlemoor Rd. 28	A12	26
Little Neville St. 1	H11	29
Little Providence St. 9	K11	30
Little Queen St. 1	H11	29
Little Russell St. Whitehall Rd.	F12	28
Little Templar La. 2 Edward St.	J11	30
Little Town La. 11	G13	37
Little Way, 17 Falkland Mt.	J6	14
Lockwood Way, 11	J14	38
Lodge La. 11	H13	37
Lodge Vw. 12	F12	28
Lombard St. 15	N11	32
Lomond Av. 18	B5	10
Lomond Pl. 7 Grosvenor Av.	J10	30
Lomond St. 7 Grosvenor Av.	J10	30
Lomond Ter. 7 Grosvenor Av.	J10	30
Londesboro' Gro. 9	L11	31
Londesboro' Ter. 9	L11	31
Long Causeway, 9	K12	30
Long Causeway, 16	F6	12
Long Close La. 9	K11	30
Longfield Av. 28	A11	26
Longfield Dr. 13	A8	18
Longfield Gro. 28	A11	26
Longfield Mt. 28	A11	26
Longfield Rd. 28	A11	26
Longfield Ter. 28	A11	26
Longley St. 10 Askern St.	K13	38
Long Meadows, 16	D2	5
Long Row, 18	C6	11
Longroyd Av. 11	J13	38
Longroyd Cres. 11	J13	38
Longroyd Gro. 11	J13	38
Longroyd Pl. 11	J13	38
Longroyd St. 11	J13	38
Longroyd St. N.11	J13	38
Longroyd Ter. 11	J13	38
Longroyd Vw. 11	J13	38
Longwood Cres. 17	L4	15

Lord Pl. 10 Belinda St.	K13	38
Lord St. 12	G12	29
Lorne Pl. 6 Woodhouse St.	H9	21
Lorraine Pl. 7	J9	22
Lorraine Ter. 7	J9	22
Lorry Bank, 7	J9	22
Louis St. 7	J9	22
Lovell Gro. 7 Lovell Rd.	J10	30
Lovell Park Gra. 7	J10	30
Lovell Park Heights, 7	J10	30
Lovell Park Hill, 7	J10	30
Lovell Park Rd. 7	J10	30
Lovell Park Towers, 7	J10	30
Lovell Pl. 7 Lovell Rd.	J10	30
Lovell Rd. 7	J10	30
Lovell St. 7	J10	30
Lovington St. 7	J10	30
Low Close St. 2 St. Mark's Rd.	H10	29
Lowell Pl. 13 Lowell Rd.	B11	26
Lowell Rd. 13	B11	26
Lowell St. 13	B11	26
Lowell Ter. 13 Lowell Rd.	B11	26
Lower Alfred St. 7 Alfred Pl.	J10	30
Lower Brunswick St. 2	J10	30
Lower Carr Pl. 10 Balm Rd.	K14	38
Lower Carr St. 10	J14	38
Lower Hanover St. 1	H11	29
Lower Town St. 13	C10	27
Lower Wortley Rd. 12	E12	28
Low Fields Rd. 12 and 11	G13	37
Low Fold, 18	B7	18
Low Fold, 9	K12	30
Low Gipton Cres. 9	M9	23
Low Grange Cres. 10	K14	38
Low Grange Vw. 10	K15	38
Low Hall Pl. 11	G12	29
Low Hall Rd. 18	A7	18
Low La. 18	C6	11
Low Mill La. 12	E13	36
Low Moor Side, 11 Towngate	H12	29
Low Moor Side La, 12	C14	35
Low Rd. 10	K13	38
Low Shops La. 26	M16	39
Lowther Cres. 26	Q14	41
Lowther Dr. 26	Q14	41
Lowther St. 8	K9	22
Lowtown, 28	A11	26
Low Wood Wk. 6	E8	20
Lucas Pl. 6	H9	21
Lucas St. 6	H9	21
Lucas Ter. 6 Institution St.	H9	21
Ludgate Hill, 2 Vicar La.	J11	30
Lulworth Av. 15	P10	33
Lulworth Cl. 15	P10	33
Lulworth Cres. 15	P10 to P11	33
Lulworth Dr. 15	P10	33
Lulworth Garth, 15	P11	33
Lulworth Vw. 15	P10	33
Lulworth Wk. 15	P10	33
Lumb Sq. 9 York St.	J11	30
Lumby La. 28	A12	26
Lumley Av. 4	F9	20
Lumley Gro. 4 Lumley Rd.	F9	20
Lumley Mt. 4 Lumley Rd.	F9	20
Lumley Pl. 4 Lumley Rd.	F9	20
Lumley Rd. 4	F9	20

Lumley St. 4 Lumley Rd.	F9	20
Lumley Ter. 4 Lumley Rd.	F9	20
Lumley Vw. 4 Lumley Rd.	F9	20
Lumley Wk. 4 Lumley Rd.	F9	20
Lunan Pl. 8	K9	22
Lunan Ter. 8	K9	22
Lupton Av. 9	L10 to L10	31
Lupton St. 10	K13	38
Luttrell Cl. 16	E6	12
Luttrell Cres. 16	E6	12
Luttrell Gdns. 16	E6	12
Luttrell Pl. 16	E6	12
Luttrell Rd. 16	E6	12
Luxor Av. 8	K9	22
Luxor Rd. 8	K9	22
Luxor St. 8	K9	22
Luxor Vw. 8	K9	22
Lydia St. 2	J11	32
Lyme Chase, 14	N10	30
Lyndhurst Rd. 14	Q8	25
Lynnfield Gdns. 14	Q8	25
Lynwood Cres. 26	P16	41
Lynwood Cres. 12	F12	28
Lynwood Gro. 12	F13	36
Lynwood Mt. 12	F12	28
Lynwood Rise, 12	F12	28
Lynwood Rd. 12	F12	28
Lynwood Vw. 12	F12	28
Lytton Ter. 10 Cariss St.	J13	38
M1 10 & 26	K14 to K17	38
Mabgate, 9 Mabgate	J11	30
Mabgate Grn. 9 Mabgate	J11	30
Macaulay St. 9	K11	30
Mafeking Av. 11 Dewsbury Rd.	G15	37
Mafeking Gro. 11 Dewsbury Rd.	G15	37
Mafeking Mt. 11 Dewsbury Rd.	G15	37
Main Pl. 10 Hunslet Rd.	J11	30
Main Rd. 12	F11	28
Main St. 14	Q8 to R8	25
Main St. 14	Q4	17
Main St. 12	F11	28
Main St. 17	N5	16
Maitland St. 11 Colville Ter.	H13	37
Malham Cl. 14	O9	24
Malmesbury Gro. 12 Amberley Rd.	F11	28
Malmesbury Pl. 12	F12	28
Malmesbury St. 12 Amberley Rd.	F11	28
Malmesbury Ter. 12	F12	28
Maltby Cl. 15	P11	33
Maltby Pl. 10 Church St.	J13	38
Malvern Pl. 11 Malvern St.	H13	37
Malvern Rd. 11	H13	37
Malvern St. 11	H13	37
Manitoba Pl. 7 Montreal Av.	J8	22
Mann Pl. 11 Domestic St.	G12	29
Mann St. 11 Domestic St.	G12	29
Manor Av. 6	G9	21
Manor Ct. 17	N5	16
Manor Dr. 6	G9	21
Manor Farm Cl. 10	J16	43
Manor Farm Cres. 27	F15	36
Manor Farm Dr. 10	J16	43
Manor Farm Dr. 27	F15	36
Manor Farm Gdns. 10	J16	43
Manor Farm Grn. 10	J16	43
Manor Farm Rd. 10	J16	43

Mitford Pl. 12 — F11 28
Mitford Rd. 12 — F11 28
Mitford Ter. 12 — F11 28
Modder Pl. 12 — E11 28
Modder Ter. 12 — E11 28
Model Av. 12 — F11 28
Model Rd. 12 — F11 28
Model Ter. 12 — F11 28
Mona Pl. 11
  Hunslet Hall Rd. — H13 37
Mona Ter. 11
  Hunslet Hall Rd. — H13 37
Mona Vw. 11
  Hunslet Hall Rd. — H13 37
Monk Bridge Av. 6
  Monk Bridge Ter. — G8 21
Monk Bridge Dr. 6 — G8 21
Monk Bridge Gro. 6
  Monk Bridge Ter. — G8 21
Monk Bridge Pl. 6
  Monk Bridge Ter. — G8 21
Monk Bridge Rd. 6 — G8 21
Monk Bridge St. 6 — G8 21
Monk Bridge Ter. 6 — G8 21
Monkswood Av. 14 — N7 24
Monkswood Bank, 14 N7 24
Monkswood Cl. 14 — N7 24
Monkswood Dr. 14 — N7 24
Monkswood Gate, 14 O7 24
Monkswood Grn. 14
  Monkswood Av. — N7 24
Monkswood Hill, 14 N7 24
Monkswood Rise, 14 N7 24
Monkswood Wk. 14 N7 24
Montagu Av. 8 L8 to M8 23
Montagu Cres. 8 — L9 23
Montagu Dr. 8 — L8 23
Montagu Gdns. 8 — L8 23
Montagu Gro. 8 — M8 23
Montagu Pl. 8 — L8 23
Montagu Vw. 8 — L8 23
Montcalm Cres. 10 — K14 38
Montpelier, 6
  Cliff Rd. — H9 21
Montreal Av. 7 — J8 22
Montreal Rd. 13 — B11 26
Moor Allerton Av. 17 K6 14
Moor Allerton Cres.
  17 — K6 14
Moor Allerton Dr. 17 K6 14
Moor Allerton Gdns.
  17 — J6 14
Moor Allerton Way,
  17 — K6 14
Moor Av. 15 — N11 32
Moor Cres. 11 — J13 38
Moor Cres. Gdns. 11 J13 38
Moor Dr. 6 — G8 21
Moorfield Av. 12 — E11 28
Moorfield Cres. 12
  Moorfield Rd. — E11 28
Moorfield Gro. 12 — E11 28
Moorfield Rd. 12 — E11 28
Moorfields, 13 — C9 19
Moorfields St. 12 — E11 28
Moor Flatts Av. 10 J16 43
Moor Flatts Rd. 10 J16 43
Moorf St. 2 — H9 21
Moor Grange Cl. 16 E7 20
Moor Grange Ct. 16 D7 19
Moor Grange Dr. 16 E7 20
Moor Grange Rise, 16 E7 20
Moor Grange Vw. 16 E7 20
Moor Gro. 28 — A12 26
Moorland Av. 27 — B15 34
Moorland Av. 6 — G10 29
Moorland Cres. 27 — B15 34
Moorland Cres. 17 — J6 14
Moorland Dr. 17 — J6 14
Moorland Gdns. 17 — J6 14
Moorland Garth, 17 — J6 14
Moorland Gro. 17 — J6 14
Moorland Ings, 17 — J6 14
Moorland Leys, 17 — H6 13
Moorland Mt. 6
  Hyde Park Rd. — G10 29
Moorland Pl. 6
  Hyde Park Rd. — G10 29

Moorland Rise, 17 — J6 14
Moorland Rd. 16 — C2 5
Moorland Rd. 6 — G10 29
Moorland Vw. 17 — J6 14
Moorland Vw. 17
  Street La. — J7 22
Moorland Walk, 17 — J6 14
Moorland Walk, 17
  Street La. — J7 22
Moor La. 16 — C2 5
Moor Park Av. 6 — F8 20
Moor Park Dr. 6 — F8 20
Moor Park Mt. 6 — F8 20
Moor Park Vils. 6 — G8 21
Moor Rd. 6 — F8 20
Moor Rd. 10 and 11 J13 38
Moorside Gro. 13
  Moorside St. — C9 19
Moorside Rd. BD11 A16 19
Moorside St. 13 — C9 19
Moorside Ter. 13
  Moorside St. — C9 19
Moortown Ring
  Rd. 17 — L5 to M6 15
Moor Vw. 11 — H12 29
Moorville Gro. 11 — H13 37
Moorville Rd. 11 — H13 37
Moresdale La.
  14 — N9 to O9 24
Morley Rd. 27 — D16 35
Morphet Gro. 7
  Upper Elmwood St. J10 30
Morphet Ter. 7
  Howarth Pl. — J10 30
Morris Av. 6 — E8 20
Morris Fold, 12
  Lower Wortley Rd. E12 28
Morris Gro. 5 — E9 20
Morris La. 6 — E8 20
Morris St. 11
  Dewsbury Rd. — G15 37
Morris Vw. 5 — E9 20
Morritt Av. 15 — O10 32
Morritt Dr. 15 — N11 32
Morritt Gro. 15 — N11 32
Moseley Pl. 6
  Speedwell St. — H9 21
Moseley St. 6 — H9 21
Moseley Ter. 6
  Crowther St. — H9 21
Moseley Wood
  App. 16 — C5 11
Moseley Wood
  Av. 16 — C4 11
Moseley Wood
  Bank, 16 — C4 11
Moseley Wood
  Cl. 16 — C5 11
Moseley Wood
  Cres. 16 — C4 11
Moseley Wood
  Croft, 16 — C4 11
Moseley Wood
  Dr. 16 — C4 11
Moseley Wood
  Gdns. 16 — C4 11
Moseley Wood
  Grn. 16 — C4 11
Moseley Wood
  La. 16 — D4 11
Moseley Wood
  Rise, 16 — C4 11
Moseley Wood
  Wk. 16 — C4 11
Moseley Wood
  Way, 16 — C4 11
Moss Gdns. 17 — H4 13
Moss Rise, 17 — H4 13
Moss Valley, 17 — H4 13
Mount Dr. 17 — H4 13
Mount Gdns. 17 — H4 13
Mount Pl. 11
  Dewsbury Rd. — G15 37
Mount Pleasant, 13 — J16 43
Mount Pleasant, 13 — B9 18
Mount Pleasant Av. 8 K8 22
Mount Pleasant Rd.
  28 — A11 26

Mount Rise, 17 — H4 13
Mount Tabor, 9
  Burmantofts St. — K11 30
Mount, The, 15 — O10 32
Mount, The, 17 — H4 13
Mulberry St. 28 — A11 26
Murton Cl. 14 — O9 24
Muschamp Ter. 7
  Benson St. — J10 30
Musgrave Bank, 13 — D10 27
Musgrave's Fold, 9
  Richmond St. — K11 30
Musgrave Mt. 13 — D10 27
Musgrave Rise, 13 — D10 27
Musgrave Vw. 13 — D10 27
Mushroom St. 9 — J10 30
Myrtle Sq. 6 — G7 21
Myrtle St. 10 — J12 30

Nancroft Cres. 12
  Brooklyn Ter. — F11 28
Nancroft Mt. 12
  Brooklyn Ter. — F11 28
Nansen Av. 13
  Station Mt. — B10 26
Nansen Gro. 13
  Station Mt. — B10 26
Nansen Mt. 13
  Station Mt. — B10 26
Nansen Pl. 13
  Station Mt. — B10 26
Nansen St. 13 — B10 26
Nansen Ter. 13
  Station Mt. — B10 26
Nansen Vw. 13
  Station Mt. — B10 26
Naseby Gdns. 9 — K10 30
Naseby Garth. 9 — K10 30
Naseby Gra. 9 — K11 30
Naseby Ter. 9 — K11 30
Naseby Vw. 9 — K11 30
Naseby Walk, 9 — K11 30
Nassau Pl. 7 — K9 22
Nayburn App. 14 — O7 24
Nayburn Cl. 14 — O7 24
Nayburn Dr. 14 — O7 24
Nayburn Pl. 14 — O7 24
Nayburn Rd. 14 — O7 24
Nayburn Vw. 14 — O7 24
Naylor St. 11
  Meynell St. — H12 29
Neath Gdns. 9 — M9 23
Neill St. 12
  Armley Rd. — F11 28
Nellie Cres. 9 — K11 30
Nellie Vw. 9 — K11 30
Neptune St. 9 — J11 30
Nesfield Cl. 10 — K16 43
Nesfield Cres. 10 — K16 43
Nesfield Gdns. 10 — K16 43
Nesfield Garth, 10 — K16 43
Nesfield Rd. 10 — K16 43
Nesfield Vw. 10 — K16 43
Nesfield Wk. 10 — K16 43
Nether Green Ct. 6
  Woodhouse St. — H9 21
Nettleton Ct. 15 — P11 33
Neville App. 9 — M12 31
Neville Av. 9 — M12 31
Neville Cl. 9 — M12 31
Neville Cres. 9 — M11 31
Neville Garth, 9 — M12 31
Neville Gro. 9 — M11 31
Neville Mt. 9 — M12 31
Neville Par. 9 — M12 31
Neville Pl. 9 — M11 31
Neville Row, 9 — M12 31
Neville St. 11 and 1 H12 29
Neville Ter. 9 — M12 31
Neville Vw. 9 — M12 31
Neville Wk. 9 — M11 31
New Adel Av. 16 — E5 12
New Adel Gdns. 16 — E5 12
New Adel La. 16 — E5 12
New Bawn, 12 — D12 27
New Briggate, 2 — J11 30

Name	Grid
New Brighton, 13	C10 27
New Gain Wk. 11	
Gain Wk.	H12 29
Newhall Bank, 10	J16 43
Newhall Chase, 10	J16 43
Newhall Cl. 10	J16 43
Newhall Cres. 10	J16 43
Newhall Croft, 10	J15 38
Newhall Garth, 10	J16 43
Newhall Grn. 10	J16 43
Newhall Mt. 10	J16 43
Newhall Rd. 10	J16 43
Newhall Wk. 10	K16 43
New Inn St. 12	E11 28
Newlaithes Gdns. 18	B7 18
Newlaithes Rd. 18	B8 18
New La. 27	B15 34
New La. 10	H16 42
Newlay Gro. 18	B8 18
Newlay La. 18	B7 18
Newlay La. 13	C9 19
Newlay La. Pl. 13	C9 19
Newlay Wood Av. 18	C7 19
Newlay Wood Cres. 18	C7 19
Newlay Wood Dr. 18	C7 19
Newlay Wood Rd. 18	B7 18
New Market St. 1	J11 30
New Park St. 1	H11 29
New Pepper Rd. 10	K13 38
Newport Cres. 6	F9 20
Newport Gdns. 6	F9 20
Newport Mt. 6	F9 20
Newport Rd. 6	F9 20
Newport Vw. 6	F9 20
New Princess St. 11	H12 29
New Ring Rd. Sea. and 14	O8 to P9 24
New Road Side, Hor.	B7 18
New Scarborough Rd. 13	C10 27
New Station St. 1	H11 29
New St. 18	B7 18
New St. 28	A12 26
New Street Gdns. 28	A12 26
New Temple Gate, 15	O12 32
Newton Cl. 7	J8 22
Newton Cres. 7	J8 22
Newton Garth, 7	K8 22
Newton Gro. 7	
Chapeltown Rd.	J9 22
Newton Lodge Dr. 7	J8 22
Newtonpark Dr. 7	K9 22
Newtonpark Vw. 7	K9 22
Newton Rd. 7	K9 22
Newton Sq. 12	C13 35
Newton Vils. 7	J8 22
New Walk, 8	L6 15
New Woodhouse La. 2	H10 29
New York Rd. 2 and 9	J11 30
New York Rd. 2	J11 30
Nice Av. 8	
Harehills La.	K9 22
Nice St. 8	K9 22
Nice Vw. 8	
Harehills Rd.	K9 22
Nickleby Rd. 9	L11 31
Nineveh Par. 11	H12 29
Nineveh Rd. 11	H12 29
Nineveh Ter. 11	
Marshall St.	H12 29
Nineveh Vw. 11	
Nineveh Rd.	H12 29
Nineveh Wk. 11	
Nineveh Rd.	H12 29
Ninth Av. 12	F12 28
Nippet La. 9	K11 30
Nixon Av. 9	L11 31
Nookin, The, 14	O9 24
Nook Rd. 14	Q7 25
Nook St. 10	
Hillidge Rd.	J13 38
Nook, The, 27	C15 35
Nook, The, 17	J4 14
Nora Pl. 13	B9 18
Nora Rd. 13	B9 18
Nora Ter. 13	B9 18
Norfolk Cl. 7	J7 22
Norfolk Gdns. 7	J7 22
Norfolk Grn. 7	J7 22
Norfolk Mt. 7	J7 22
Norfolk Pl. 7	J7 22
Norfolk St. 10	J12 30
Norfolk Ter. 7	
Pasture La.	J7 22
Norfolk Vw. 7	J7 22
Norfolk Wk. 7	J7 22
Norman Gro. 5	E9 20
Norman Mt. 5	E9 20
Norman Pl. 8	L6 15
Norman Row, 5	E9 20
Norman St. 5	E9 20
Norman Ter. 8	L6 15
Normanton Gro. 11	
Cambrian Rd.	H13 37
Normanton Pl. 11	
Cambrian Rd.	H13 37
Normanton St. 11	H13 37
Normanton Ter. 11	
Cambrian Rd.	H13 37
Normanton Vw. 11	
Cambrian Rd.	H13 37
Norman Towers, 16	E7 20
Norman Vw. 5	E9 20
North Broadgate La. 18	C6 11
Northbrook Pl. 7	J7 22
Northbrook St. 7	J7 22
North Cl. 8	M8 23
North Farm Rd. 8	L7 to M9 23
North Grange Mt. 6	G8 21
North Grange Rd. 6	G9 21
North Grove Cl. 8	M8 23
North Grove Dr. 8	M8 23
North Grove Rise, 8	M8 23
North Hall St. 3	
Burley Rd.	F10 28
North Hall Ter. 3	
Newton St.	G11 29
North Hill Rd. 6	G9 21
North La. 8	M7 23
North La. 6	F9 20
North Lingwell Rd. 10	J16 43
North Mead 16	D2 5
Northolme Av. 16	E7 20
Northolme Cres. 16	E7 20
North Parade, 16	E7 20
North Park Av. 8	K7 22
North Park Gro. 8	L7 to L6 23
North Park Par. 8	K6 14
North Park Rd. 8	L7 23
North Parkway, 14	N8 to O9 24
North Pl. 7	
North St.	J10 30
North Rd. 18	B5 10
North Rd. 15	P10 33
North St. 2 and 7	J10 30
North Ter. 15	P10 33
North View St. 28	
Arthur St.	A10 26
North View Ter. 28	
Arthur St.	A10 26
North Way, 8	M8 23
North West Gro. 6	H9 21
North West Pl. 6	H9 21
North West Rd. 6	H9 21
North West St. 6	H9 21
North West Ter. 6	H9 21
North West Vw. 6	H9 21
Northwood Cl. 26	P15 41
Northwood Falls, 26	P15 41
Northwood Mt. 28	A12 26
Northwood Park, 26	P15 41
Northwood Vw. 28	A12 26
Norton Rd. 8	L5 15
Norwich Av. 10	J14 38
Norwich Pl. 10	J14 38
Norwich Row, 10	
Norwich Av.	J14 38
Norwich St. 10	
Norwich Av.	J14 38
Norwood Gro. 6	G9 21
Norwood Mt. 6	G9 21
Norwood Pl. 6	G9 21
Norwood Rd. 6	G9 21
Norwood Ter. 6	G9 21
Noster Gro. 11	G13 37
Noster Pl. 11	G13 37
Noster Rd. 11	G13 37
Noster St. 11	G13 37
Noster Ter. 11	G13 37
Noster Vw. 11	
Noster Rd.	G13 37
Nowell Av. 9	L10 31
Nowell Cres. 9	L10 31
Nowell End Row, 9	L10 31
Nowell Gdns. 9	L10 31
Nowell Gro. 9	L10 31
Nowell La. 9	L10 31
Nowell Mt. 9	L10 31
Nowell Par. 9	L10 31
Nowell Pl. 9	L10 31
Nowell St. 9	L10 31
Nowell Ter. 9	L10 31
Nowell Vw. 9	L10 31
Nowell Way, 9	L10 31
Nunnington Av. 12	
Armley Park Rd.	F10 28
Nunnington St. 12	F11 28
Nunnington Ter. 12	
Armley Park Rd.	F10 28
Nunroyd Av. 17	J6 14
Nunroyd Gro. 17	J6 14
Nunroyd Lawn, 17	J6 14
Nunroyd Rd. 17	J6 14
Nunroyd St. 17	J6 14
Nunroyd Ter. 17	J6 14
Nunthorpe Rd. 13	A8 18
Nursery Cl. 17	J5 14
Nursery Gro. 17	H5 13
Nursery La. 17	H5 13
Nursery Mt. 10	
Nursery Mt. Rd.	K14 38
Nursery Mt. Rd. 10	K14 38
Nutting Gro. 12	
Cross La.	E11 28
Oak Cres. 15	N11 32
Oakdene Cl. 28	A12 26
Oakfield St. 7	
Cambridge Rd.	H9 21
Oakfield Ter. 6	G8 21
Oakhurst Av. 11	H14 37
Oakhurst Gro. 11	G14 37
Oakhurst Mt. 11	G14 37
Oakhurst Rd. 11	G14 37
Oaklands Av. 13	A8 18
Oaklands Cl. 13	A8 18
Oaklands Rd. 13	A8 18
Oaklea Gdns. 16	F6 12
Oaklea Rd. 14	Q8 25
Oakley Gro. 11	J13 38
Oakley Ter. 11	J14 38
Oak Rd. 7	J8 22
Oak Rd. 12	G11 29
Oak Rd. 15	N11 32
Oakroyd Mt. 28	A11 26
Oakroyd Ter. 28	A11 26
Oak St. 27	E15 36
Oak Tree Cl. 9	M9 23
Oak Tree Cres. 9	M9 23
Oak Tree Dr. 8	M9 23
Oak Tree Gro. 9	M9 23
Oak Tree Mt. 9	M9 23
Oak Tree Pl. 9	M9 23
Oak Tree Wk. 9	M9 23
Oak Villas, 17	K5 14
Oakwell Av. 8	L8 23
Oakwell Cres. 8	L8 23
Oakwell Dr. 8	L8 23

Park Wood Dr. 11 G15 37
Park Wood Rd. 11 G15 37
Parliament Pl. 12
  Armley Rd. F11 28
Parliament Rd. 12 F11 28
Parliament St. 1
  Grace St. H11 29
Parliament Ter. 12 F11 28
Parnaby Cres. 10 K14 38
Parnaby Mt. 10 K14 38
Parnaby Rd. 10 K14 38
Parnaby Ter. 10
  Parnaby Rd. K14 38
Parnaby Vw. 10 K14 38
Parsonage St. 4 F10 28
Parson St. 10 K12 30
Pasture Av. 7 J7 22
Pasture Cres. 7
  Pasture Av. J7 22
Pasture Gro. 7
  Pasture La. J7 22
Pasture La. 7 J7 22
Pasture Mt. 12 E11 28
Pasture Par. 7
  Pasture Av. J7 22
Pasture Pl. 7
  Pasture Av. J7 22
Pasture Rd. 8 K9 22
Pasture St. 7
  Pasture Av. J7 22
Pasture Ter. 7
  Pasture Av. J7 22
Pasture Vw. 12 E11 28
Patchetts La. 13 C9 19
Pearson Av. 6
  Brudenell Rd. G9 21
Pearson Gro. 6 G9 21
Pearson St. 10 J12 30
Pearson Ter. 6 G9 21
Peel Ct. 6
  Woodhouse St. H9 21
Peel St. 28 A12 26
Pembroke Gra. 9 M10 31
Pembroke Rd. 28 A11 26
Pembroke Towers, 9 M10 31
Penda's Dr. 15 P10 33
Penda's Gro 15 P9 25
Penda's Wk. 15 P10 33
Penda's Way, 15 P10 33
Pennington Ct. 6
  Woodhouse St. H9 21
Pennington Pl. 6 H9 21
Pennington St. 6 H9 21
Pennington Ter. 6 H9 21
Penraevon Av. 7 19 22
Penraevon Pl. 7
  Meanwood Rd. G8 21
Penraevon St. 7 J9 22
Penraevon Ter. 7
  Meanwood Rd. G8 21
Penrith Gro. 12 F12 28
Penwell Dean, 14 P8 25
Penwell Fold, 14 P8 25
Penwell Garth, 14 P8 25
Penwell Gate, 14 P8 25
Penwell Grn. 14 P8 25
Penwell Lawn, 14 P8 25
Pen-y-Ffynnon Rd. 6
  Parkside Rd. G6 13
Pepper Gro. 10 K13 38
Pepper La. 10 K13 38
Pepper La. 9 C9 19
Pepper Pl. 10 K13 38
Pepper Rd. 10 K14 38
Pepper St. 10 K13 38
Pepper Ter. 10 K13 38
Pepper Vw. 10
  Pepper La. K13 38
Percival St. 2
  Cookridge St. H11 29
Perseverance St. 11
  Sydenham St. G12 29
Perseverance St. 10 L14 39
Perth Mt. 18 B5 10
Peter La. 27 G16 42
Petrie Cres. 13 A8 18
Petrie St 13 A8 18
Pickard Ct. 15 P11 33

Pickard Pl. 12
  Campbell St. G11 29
Pickering Mt. 12
  Pickering St. F11 28
Pickering Pl. 12
  Pickering St. F11 28
Pickering St. 12 F11 28
Pickering Ter. 12
  Pickering St. F11 28
Pickpocket La.
  26 O15 40
Piecewood Rd. 16 C5 11
Pigeon Cote Cl. 14 O8 24
Pigeon Cote Rd. 14 N8 24
Pilot St. 9 K10 30
Pinder Av. 12 D13 35
Pinder St. 12 D13 35
Pinder Vw. 12 D13 35
Pinfold Ct. 15 O11 32
Pinfold Gro. 15 O11 32
Pinfold Hill, 15 O11 32
Pinfold La. 12 E11 28
Pinfold La. 15 O11 32
Pinfold La. 16 D4 11
Pinfold Mt. 15 O11 32
Pinfold Rd. 15 O11 32
Pink Hill, 12 E11 28
Pipe and Nook La. 12 D11 27
Pitfall St. 2
  Call La. J11 30
Place's Rd. 9 K11 30
Place, The, 8 L7 23
Plantation Av. 15 N11 32
Plantation Av. 17 L4 15
Plantation Gdns. 17 L4 15
Playfair Av. 10
  Royal Rd. J14 38
Playfair Cres. 10
  Royal Rd. J14 38
Playfair Gro. 10 J14 38
Playfair Rd. 10 J14 38
Playfair St. 10 J14 38
Playfair Ter. 10 J14 38
Playfair Vw. 10 J13 38
Playground, 12 C13 35
Pleasant Av. 11
  Pleasant St. G12 29
Pleasant Grn. 6
  Rampart Rd. H9 21
Pleasant Gro. 11
  Pleasant St. G12 29
Pleasant Mt. 11
  Domestic St. G12 29
Pleasant Pl. 11 H12 29
Pleasant Rd. 11
  Pleasant St. G12 29
Pleasant St. 11 G12 29
Pleasant Ter. 11
  Domestic St. G12 29
Pleasant Vw. 11 H12 29
Plevna St. 10 L14 39
Plevna Ter. 10
  Grove Rd. K13 38
Plum St. 2
  Gower St. J11 30
Poets Pl. 18 C6 11
Pollard St. 28
  Leeds&Bradford Rd. A10 18
Pollard St. 13 C8 19
Pollard Pl. 12
  Lord St. G12 29
Pollard St. 12 G12 29
Pontefract Av. 9 K11 30
Pontefract Gro. 9
  Pontefract St. K11 30
Pontefract La., 9, 15
  K11 to Q14 30
Pontefract Rd. 10, & 26
  L14 to O16 39
Pontefract St. 9 K11 30
Pontefract Ter. 12 D12 27
Poole Cres. 15 O10 32
Poole Mt. 15 O10 32
Poole Rd. 15 O10 32
Poole Sq. 15 O10 32
Poplar Av. 15 N11 32
Poplar Av. 15 P10 33
Poplar Ct. 13 D11 27

Poplar Croft, 13 D11 27
Poplar Dr. 18 A6 10
Poplar Gdns. 13 D11 27
Poplar Garth 13 D11 27
Poplar Gro. 4
  Poplar St. G10 29
Poplar Mt. 13 D11 27
Poplar Rise, 13 D10 27
Poplar Rd. 4
  Burley Rd. F10 28
Poplars, The, 6 G9 21
Poplars, The, 16 D2 5
Poplar Vw. 13 D10 27
Poplar Way, 13 D11 27
Portage Av. 15 N11 32
Portage Cres. 15
  Portage Av. N11 32
Portland Cres. 1 H11 29
Portland Gate, 1
  Portland Cres. H11 29
Portland Rd. 12 F12 28
Portland St. 1 H11 29
Portland Way, 1 H10 29
Porto St. 11
  Moor Cres. J13 38
Potternewton Av. 7 H8 21
Potternewton Cres. 7 H8 21
Potternewton Gro. 7 H8 21
Potternewton Heights,
  7 J8 22
Potternewton La. 7 H8 21
Potternewton Mt. 7 H8 21
Potternewton Vw. 7 H8 21
Pottery La. 26 P15 41
Pottery Vale, 10 J13 38
Potter Row, 11
  Jack La. H12 29
Poulton Pl. 11 J13 38
Preston Par. 11 H14 38
Pretoria Av. 9 K12 30
Pretoria Cl. 9 K12 30
Pretoria Gro. 9 K12 30
Pretoria Mt. 9 K12 30
Pretoria Pl. 9 K12 30
Pretoria Rd. 9 K12 30
Pretoria St. 9 K12 30
Pretoria Ter. 9 K12 30
Pretoria Vw. 9 K12 30
Priestley Dr. 28 B10 26
Priestley Gdns. 28 A11 26
Priestley St. 28 B11 26
Priestley Wk. 28 B10 26
Primley Gdns. 17 J5 14
Primley Park Av. 17 J5 14
Primley Park Ct. 17 J4 14
Primley Park Cres. 17 J5 14
Primley Park Cres. E.
  17 J5 14
Primley Park Cres. W.
  17 J5 14
Primley Park Garth, 17 J4 14
Primley Park Grn. 17 J4 14
Primley Park Gro. 17 J5 14
Primley Park La. 17 J5 14
Primley Park Mt. 17 J5 14
Primley Park Rise, 17 J5 14
Primley Park Rd. 17 J5 14
Primley Park Vw. 17 J4 14
Primley Park Wk. 17 J4 14
Primley Park Way, 17 J4 14
Primrose Av. 15 O11 32
Primrose Cl. 15 O11 32
Primrose Cres. 15 O10 32
Primrose Dr. 15 O11 32
Primrose Gdns. 15 O10 32
Primrose Garth, 15 N11 32
Primrose Gro. 15 O11 32
Primrose Hill, 28 A10 26
Primrose Hill, 12 D12 27
Primrose Hill, 15 O11 32
Primrose Hill, 28
  Richardshaw La. A10 26
Primrose La. 11 J13 38
Primrose La. 15 N11 32
Primrose Rd. 15 O11 32
Primrose Wk. 27 F15 36
Prince Augustus St. 12
  Langham St. G11 29

Prince Edward Gro.
12 E13 36
Prince Edward Rd. 12 E13 36
Prince's Av. 8 L7 23
Princess Field Pl. 11 H12 29
Privilege St. 12 E11 28
Proctors Pl. 7 J9 22
Prospect Av. 9
Prospect St. J13 38
Pospect Pl. 18 B7 18
Prospect Pl. 13 A8 18
Prospect Pl. 13
Prospect St. C9 19
Prospect Rise, 17 N5 16
Prospect St. 10 L14 39
Prospect St. 13 C9 19
Prospect Ter. BD11 A15 34
Prospect Ter. 10 J13 38
ospect Vw. 15 P10 33
Prospect Vw. 13
Upper Town St. C9 19
Prosper Ter. 10
PrJoseph St. J13 38
Prosperity St. 7 J10 30
Providence Av. 6 H9 21
Providence Rd. 6 H9 21
Providence St. 9 K11 30
Providence Ter. 2
Raglan Rd. H9 21
Provost St. 11 H12 29
Pudsey Rd. 28.
13 and 12 B11 to D11 26
Pulleyn St. 10 J12 30

Quarry Bank Ct. 5 D8 19
Quarry Gdns. 17 H4 13
Quarry Mt. 6
Delph La. H9 21
Quarry Mt. Pl. 6
Lucas St. H9 21
Quarry Mt. St. 6
Delph La. H9 21
Quarry Mt. Ter. 7
Lucas St. H9 21
Quarry Pl. 6 H9 21
Quarry St. 6 H9 21
Quarry Ter. 18 B6 10
Quarry, The, 17 H4 13
Quarry Vw. Ter. 13
Hough End La. C10 27
Quebec St. 1 H11 29
Queen's Arc. 1
Briggate J11 30
Queen's Ct. 1
Briggate J11 30
Queenshill App. 17 J6 14
Queenshill Av. 17 J6 14
Queenshill Cres. 17 J5 14
Queenshill Dr. 17 H6 13
Queenshill Gdns. 17 H6 13
Queenshill Garth, 17 J6 14
Queenshill Rd. 17 J6 14
Queenshill Vw. 17 J6 14
Queenshill Wk. 17 J6 14
Queens Par. 14 O8 24
Queen's Pl. 2 J10 30
Queen Sq. 2 J10 30
Queen Sq. 2
Woodhouse La. G9 21
Queen's Rd. 6 G10 29
Queensthorpe Av.
13 C11 27
Queensthorpe Cl. 13 D11 27
Queensthorpe Rise,
13 C11 27
Queen St. 10 L14 39
Queen St. 1 H11 29
Queensview, 14 O8 24
Queensway, 15 O11 32
Queenswood Cl. 6 E8 20
Queenswood Dr. 6
E7 to F9 20
Queenswood Gdns. 6 F9 20
Queenswood Grn. 6 E7 20
Queenswood Mt. 6 E8 20
Queenswood Rise, 6 E9 20

Queenswood Rd. 6 E8 20
Queen Victoria St. 1 J11 30

Raby St. 7 J9 22
Radcliffe Gdns. 28 A12 26
Radcliffe La. 28 A11 26
Radcliffe Ter. 28 A12 26
Rae St. 7 J9 22
Raglan Pl. 6
Woodhouse St. H9 21
Raglan Rd. 6 and 2 H9 12
Raglan Ter. 2
Raglan Rd. H9 21
Railsford Mt. 13 C10 27
Railway St. 9 K11 30
Raincliffe Gro. 9 L11 31
Raincliffe Rd. 9 L11 31
Raincliffe St. 9 L11 31
Raincliffe Ter. 9
Raincliffe Rd. L11 31
Rainville Av. 13 D9 19
Rainville Gro. 13 D9 19
Rainville Mt. 13 D9 19
Rainville Ter. 13 D9 19
Rakehill Rd. Scholes Q7
Rampart Rd. 6 H9 21
Ramsden Ter. 7 J10 30
Ramsey St. 12
Parliament Rd. F11 28
Ramshead App. 14 O8 24
Ramshead Cl. 14 N7 24
Ramshead Dr. 14 N7 24
Ramshead Gdns. 14 N7 24
Ramshead Heights, 14 O8 24
Ramshead Hill, 14 N8 24
Ramshead Pl. 14 O8 24
Randolph St. 13 B10 26
Rathmell Rd. 15 N11 32
Raven Rd. 6 G9 21
Ravenscar Av. 8 L8 23
Ravenscar Mt. 8 L8 23
Ravenscar Pl. 8 L8 23
Ravenscar Ter. 8 L8 23
Ravenscar Vw. 8 L8 23
Ravenscar Wk. 8 L8 23
Ravens Mt. 28 A11 26
Rawdon Rd. 18 A6 10
Rawson Ter. 11 J13 38
Rawson Ter. 11
Dewsbury Rd. G15 37
Raylands Cl. 10 K16 43
Raylands Garth, 10 K16 43
Raylands La. 10 K16 43
Raylands Pl, 10 K16 43
Raylands Rd. 10 K16 43
Raylands Way, 10 K16 43
Raynel App. 16 E5 12
Raynel Cl. 16 E5 12
Raynel Dr. 16 E5 12
Raynel Gdns. 16 E5 12
Raynel Grn. 16 E5 12
Raynel Mt. 16 E5 12
Raynel Way, 16 D5 11
Rayner Ter. 28 A11 26
Raynville App. 13 D10 27
Raynville Ct. 13 D10 27
Raynville Cres. 12 D10 27
Raynville Dr. 13 D9 19
Raynville Grange, 13 D9 19
Raynville Grn. 13 D10 27
Raynville Rise, 13 D10 27
Raynville Rd. 13 C9 19
Raynville Wk. 13 D10 27
Recorder St. 9
Beckett St. K10 30
Recreation Av. 11
Elland Rd. G13 37
Recreation Gro. 11 G13 37
Recreation Mt. 11 G13 37
Recreation Pl. 11 G13 37
Recreation Rd. 11 G13 37
Recreation St. 11 G13 37
Recreation Ter. 11 G13 37
Recreation Vw. 11 G13 37

Rectory St. 9 K10 30
Red Cote, 12
Stanningley Rd. D10 27
Redcote La. 12 and 4 E10 28
Redesdale Gdns. 16 E5 12
Red Hall App. 14 O6 16
Red Hall Av. 17 N6 16
Red Hall Chase, 14 O6 16
Redhall Cl. 11 G14 37
Red Hall Ct. 14 O6 16
Redhall Cres. 11 G14 37
Red Hall Croft, 14 O7 24
Red Hall Dr. 14 O6 16
Red Hall Gdns. 17 N6 16
Redhall Garth, 11 G14 37
Red Hall Garth, 14 O6 16
Red Hall Grn. 14 O6 16
Red Hall La. 17 and 14 N6 16
Red Hall Vale, 14 O7 24
Red Hall Wk. 14 O6 16
Red Hall Way, 14 O6 16
Redmire Dr. 14 O9 24
Redmire Vw. 14 O9 24
Redshaw Mt. 12
Hartington Rd. F12 28
Redshaw Rd. 12 F12 28
Redshaw St. 12
Hartington Rd. F12 28
Redshaw Vw. 12
Hartington Rd. F12 28
Regent Av. 18 C7 19
Regent Park Av. 6 G9 21
Regent Park Ter. 6 G9 21
Regent Rd. 18 B7 18
Regent St. 7 J7 22
Regent St. 2 J11 30
Regent Ter. 6 G10 29
Regina Dr. 7 J8 22
Reginald Mt. 7 J9 22
Reginald Pl. 7
Reginald Mt. J9 22
Reginald Row, 7 J9 22
Reginald St. 7 J9 22
Reginald Ter. 7 J9 22
Reginald Vw. 7
Reginald Mt. J9 22
kehoboth Rd. 12
Lower Wortley Rd. E12 28
Rein Rd. 18 B8 18
Rein, The, 14 N8 24
Reinwood Av. 8
M8 to N8 24
Renfield Pl. 7
Manor St. J10 30
Renfield Ter, 7
Manor St. J10 30
Rhoda St. 12
Amberley Rd. F11 28
Rhodes Fl. 12
Wellington Rd. G12 29
Rhodes Sq. 7
Meanwood Rd. G8 21
Rhodes St. 11 H12 29
Richardshaw Dr.
28 A10 26
Richardshaw La.
28 A10 26
Richardshaw Rd.
28 A10 26
Richardson Cres. 9 L11 31
Richardson Rd. 9 L11 31
Richard St. 10 K13 38
Richmond Av. 6 G9 21
Richmond Cl. 26 N16 40
Richmond Ct. 26
N16 40
Richmond Grn. St. 9 K11 30
Richmond Mt. 6 G9 21
Richmond Rd. 28 B11 26
Richmond Rd. 6 G9 21
Richmond St. 9 K11 30
Richmond Ter. 28 B11 26
Richmond Vw. 12
Wesley Rd. F11 28
Rickard St. 12
Whitehall Rd. F12 28

69

Rider Rd. 6	H9	21	Rock St. 9			Rosedale Wk. 10	J14	38
Rider St. 9	K11	30	*Nippet La.*	K11	30	Rosemont Av. 13	C10	27
Ridge Cl. 8	K8	22	Roderick St. 12			Rosemont Gro. 13		
Ridge End Vils, 6			*Privilege St.*	E11	28	*Rosemont Av.*	C10	27
Wood La.	F8	20	Rodley Fold, 13	*		Rosemont Pl. 13		
Ridge Gro. 7	H9	21	*Town St.*	A8	18	*Station Mt.*	B10	26
Ridge Mt. 6			Rodley La. 13	A8	18	Rosemont Rd. 13	C10	27
Cliff Rd.	H9	21	Roger Pl. 7			Rosemont St. 13	C10	27
Ridge Mount Ter. 6	H9	21	*Skinner La.*	J10	30	Rosemont Ter. 13		
Ridge Rd. 7	H9	21	Roger Row, 11	G14	37	*Station Mt.*	B10	26
Ridge Ter. 6	G8	21	Rokeby Gdns. 6	F8	20	Rosemont Vw. 13		
Ridge Way, 8	K8	22	Roker Av. 28	A12	26	*Rosemont Av.*	C10	27
Rigton App. 9	K11	30	Roker Cres. 28	A12	26	Rosemont Wk. 13	C10	27
Rigton Cl. 9	K11	30	Roker Dr. 28	A12	26	Rosemount Av. 28	A11	26
Rigton Dr. 9	K11	30	Roker Gro. 28	A12	26	Rosemount Ter.		
Rigton Grn. 9	K11	30	Roker La. 28	A12	26	28	A11	26
Rigton Lawn, 9	K11	30	Roker Mt. 28	A12	26	Rose Mount Ter. 12		
Rillbank Gro. 3	G10	29	Roker Vw. 28	A12	26	*Redshaw Rd.*	F12	28
Rillbank Pl. 3			Roman Av. 8 · L5 to L6		15	Roseneath Pl. 12	F12	28
Rillbank Rd.	G10	29	Roman Cres. 8	L6	15	Roseneath St. 12	F12	28
Rillbank Rd. 3	G10	29	Roman Dr. 8	L6	15	Roseneath Ter. 12	F12	28
Rillbank St. 3	G10	29	Roman Gdns. 8	L6	15	Rose St. 18	B7	18
Rillbank Ter. 3			Roman Gra. 8	L6	15	Rose Ter. 18	B7	18
Rillbank Rd.	G10	29	Roman Mt. 8			Roseville Av. 8		
Rillbank Vw. 3			*Roman Vw.*	L6	15	*Roseville Rd.*	K10	30
Rillbank Rd.	G10	29	Roman Pl. 8	L6	15	Roseville Rd. 8	K10	30
Ring Rd. Beeston,			Roman St. 8	L6	15	Roseville St. 8	K10	30
11	H16	42	Roman Ter. 8	L6	15	Roseville Ter. 8	K10	30
Ring Rd. Beeston,			Roman Vw. 8	L6	15	Roseville Ter. 15	P9	25
12 and 11	F13	36	Rombalds Av. 12			Rosgill Dr. 14	N8	24
Ring Rd. Beeston Pk.			*Armley Lodge Rd.*	F10	28	Rosgill Grn. 14	O8	24
11	G15	37	Rombalds Cres. 12			Rosgill Wk. 14	N8	24
Ring Rd. Bramley	B10	26	*Armley Lodge Rd.*	F10	28	Rossal Rd. 8	K9	22
Ring Rd. Churwell,			Rombalds Gro. 12			Rossefield App. 13	C10	27
11	G14	37	*Armley Lodge Rd.*	F10	28	Rossefield Dr. 13	C10	27
Ring Rd. Cross Gates,			Rombalds Pl. 12			Rossefield Gro. 13	C10	27
15	P9	25	*Armley Lodge Rd.*	F10	28	Rossefield Par. 13	C10	27
Ring Rd. Farnley,			Rombalds St. 12			Rossefield Wk. 13	C10	27
12	C11	27	*Armley Lodge Rd.*	F10	28	Rossefield Way, 13	C10	27
Ring Rd. Halton, 15	P10	33	Rombalds Ter. 12			Ross Gro. 13	B9	18
Ring Rd. Inner			*Armley Lodge Rd.*	F10	28	Rossington Gro. 8	K9	22
1 & 2	H11 to J11	29	Rombalds Vw. 12			Rossington Pl. 8		
Ring Rd. Low Wortley,			*Armley Lodge Rd.*	F10	28	*Gathorne Ter.*	K9	22
12	D12	27	Rookwood Av. 9	M11	31	Rossington Rd. 8	L8	23
Ring Rd. Meanwood,			Rookwood Cres. 9	M11	31	Rossington St. 2	H11	29
16 and 6	F6	12	Rookwood Croft, 9	M11	31	Rothbury Gdns. 16	E5	12
Ring Road, Middleton,			Rookwood Gdns. 9	M11	31	Rothsay Mt. 11		
10	K16	43	Rookwood Gro. 9	M11	31	*Little Town La.*	G13	37
Ring Rd. Moortown,			Rookwood Hill, 9	M11	31	Rothsay Pl. 11		
6 and 17	H6 to L5	13	Rookwood Mt. 9	M11	31	*Elland Rd.*	G13	37
Ring Rd. Weetwood,			Rookwood Par. 9	M11	31	Rothsay St. 11		
16	E6	12	Rookwood Pl. 9	M11	31	*Elland Rd.*	G13	37
Ring Rd. West Park,			Rookwood Rd. 9	M11	31	Rothsay Ter. 11		
16	D7	19	Rookwood Sq. 9	M11	31	*Little Town La.*	G13	37
Ringrose St. 11			Rookwood St. 9	M11	31	Rothsay Vw. 11		
Domestic St.	G12	29	Rookwood Ter. 9	M11	31	*Little Town La.*	G13	37
Ringwood Av. 14	N7	24	Rookwood Vale, 9	M11	31	Roundham Ct. 8	M7	23
Ringwood Cres. 14	O6	16	Rookwood Vw. 9	M11	31	Roundhay Av. 8	K8	22
Ringwood Dr. 14	O7	24	Rooms La. 27	D15	35	Roundhay Cres. 8	K8	22
Ringwood Gdns. 14	O7	24	Roper Av. 8	K7	22	Roundhay Gdns. 8	K8	22
Ringwood Mt. 14	N7	24	Roper Gro. 8	K7	22	Roundhay Gro. 8	K8	22
Ringwood Par. 14	N6	16	Roscoe St. 7			Roundhay Mt. 8	K9	22
Ripley Mt. 28	A10	26	*Roundhay Rd.*	K9	22	Roundhay Park La. 17	L5	15
Ripley Pl. 28	A10	26	Roscoe Ter. 7			Roundhay Pk. Rd. 17	L5	15
Ripley St. 28	A10	26	*Chapeltown Rd.*	J10	30	Roundhay Pl. 8	K8	22
Ripley Ter. 28	A10	26	Roscoe Ter. 12	E11	28	Roundhay Rd.		
Ripley Yd. 11			Rose Av. 18	B7	18	7 and 8 J10 to L8		30
Sydenham St.	G12	29	Rosebank Cres. 1			Roundhay St. 7	J10	30
Rise, The, 6	E8	20	*Rosebank Rd.*	G10	29	Roundhay Ter. 7		
Ritter St. 2	H10	29	Rosebank Gro. 1	G10	29	*Roundhay Rd.*	J10	30
Riviera Gdns. 7	J8	22	Rosebank Pl. 3			Roundhay Vw. 8	K8	22
Robb Av. 11	H14	37	*Westfield Cres.*	G10	29	Rowans, The, 16	D2	5
Robb St. 11	H14	37	Rosebank Rd. 3	G10	29	Rowland Pl. 11	H13	37
Roberts Av. 9	L10	31	Rosebank Row, 3			Rowland Rd. 11	H13	37
Roberts Gro. 9	L10	31	*Rosebank Rd.*	G10	29	Rowland Ter. 11		
Robert St. 11			Rosebank St. 3			*Dewsbury Rd.*	G15	37
Sweet St.	H12	29	*Westfield Cres.*	G10	29	Roxby Cl. 9	K10	30
Robin La. 28	A11	26	Rosebank Vw. 3	G10	29	Roxholme Av. 7	K8	22
Rob St. 28			Roseberry Ter. 28			Roxholme Gro. 7	K8	22
Cavendish Pl.	A10	26	*Gladstone Ter.*	A10	26	Roxholme Pl. 7	K8	22
Rocheford Gro. 10	K13	38	Rosebud Wk. 8	K10	30	Roxholme Rd. 7	K8	22
Rocheford Pl. 10	K13	38	Rosecliffe Mt. 13			Roxholme Ter. 7	K8	22
Rocheford Rd. 10			*Ferncliffe Rd.*	C10	27	Royal Av. 10	J14	38
New Pepper Rd.	K13	38	Rosecliffe Ter. 13			Royal Gro. 10	J14	38
Rocheford Ter. 10	K13	38	*Ferncliffe Rd.*	C10	27	Royal Park Av. 6	G10	29
Rochester Ter. 6	F9	20	Rosedale Bank, 10	J14	38	Royal Park Gro. 6	G10	29
Rockery Rd. 18	C6	11	Rosedale Gdns. 10	J14	38	Royal Park Rd. 6	G10	29
Rock Lane, 13	B9	18	Rosedale Gro. 10	J14	38	Royal Park Ter. 6	G10	29

Royal Park Vw. 6
Royal Park Gro. G10 29
Royal Pl. 10 J14 38
Royal Rd. 10 J14 38
Royal St. 10
Norwich Pl. J14 38
Royal Ter. 10
Royal Rd. J14 38
Royds La. 12 F13 36
Royston Pl. 6
Woodhouse St. H9 21
Ruby St. 9 J10 30
Rufus St. 10
Orchard St. K12 30
Runswick Av. 11 G12 29
Runswick Pl. 11 H12 29
Runswick St. 11 G12 29
Runswick Ter. 11 H12 29
Russell St. 1 H11 29
Ruth St. 12
Mickley St. F11 28
Ruth Ter. 12
Mickley St. F11 28
Ruthven Vw. 8 L9 23
Rycroft Av. 13 B10 26
Rycroft Cl. 13 B10 26
Rycroft Ct. 13 B10 26
Rycroft Dr. 13 B10 26
Rycroft Grn. 13 B10 26
Rycroft Pl. 13 B10 26
Rycroft Sq. 13 B10 26
Rycroft Towers, 13 B10 28
Rydall Pl. 11 G12 29
Rydall St. 11 G12 29
Rydall Ter. 11 G12 29
Rydedale Av. 12 E13 36
Ryder Gdns. 8 L7 23
Rye Pl. 14 N10 32
Rylstone Gro. 10
Jack La. H12 29
Rylstone Lawn, 10 J13 38
Rylstone Pl. 10
Bleasby St. J13 38
Rylstone St. 10
Bleasby St. J13 38
Rylstone Ter. 10
Jack La. H12 29

Sackville St. 7 J9 22
Sackville Ter. 7 J10 30
Sagar Pl. 6 H9 21
Sagar Rd. 6 H9 21
Sagar Musgrave's Pl. 13
Out Gang C9 19
Sagar Musgrave's Row,
13
Out Gang · C9 19
St. Agnes Gro. 9
Beckett St. K10 30
St. Agnes Mt. 9 K10 30
St. Agnes Ter. 9 K10 30
St. Alban App. 9 M10 31
St. Alban Cl. 9 M10 31
St. Alban Cres. 9 M10 31
St. Alban Gro. 9 M10 31
St. Alban Mt. 9 M10 31
St. Alban Rd. 9 M10 31
St. Albans Pl. 2
Wade La. J11 30
St. Albans St. 2 J10 30
St. Alban Vw. 9 M10 31
St. Andrews Cl. 13 A8 18
St. Andrews Pl. 3
Darlington St. G11 29
St. Anne's Dr. 4 F9 20
St. Anne's Gdns. 4 F9 20
St. Anne's Grn. 4 F9 20
St. Anne's La. 4 F9 20
St. Anne's Rd. 6 F8 20
St. Ann's Av. 4 F10 28
St. Ann's Cl. 4 F9 20
St. Ann's Mt. 4 F9 20
St. Ann's Rise, 4 E9 20
St. Ann's Sq. 4 F9 20
St. Ann's Way, 4 F9 20
St. Ann St. 2 H11 29

St. Anthony's Dr. 11
Back La. G14 37
St. Anthony's Rd. 11 G14 37
St. Barnabas' Rd. 11 H12 29
St. Barnabas' St. 11
Sweet St. H12 29
St. Barnabas' Ter..11
Sweet St. H12 29
St. Catherine's Cres.
13 C9 19
St. Catherine's Dr.
13 C9 19
St. Catherine's Grn.
13 C8 19
St. Catherine's Hill,
13 C9 19
St. Catherine's Wk. 8 L8 23
St. Chad's Av. 6 F8 20
St. Chad's Dr. 6 F8 20
St. Chad's Gro. 6 F8 20
St. Chad's Rise, 6 F8 20
St. Chad's Rd. 16 F8 20
St. Chad's Vw. 6 F8 20
St. Columba St. 2
Woodhouse La. G10 29
St. Elmo Gro. 9 L11 31
St. Francis Pl. 11
· *Back Row* H12 29
St. George's Av.
26 M15 39
St. Helens Cl. 16 F5 12
St. Helens La. 16 E5 12
St. Helen's St. 10 J12 30
St. Hilda's Av. 9 K12 30
St. Hilda's Ct. 9 K12 30
St. Hilda's Gro. 9 K12 30
St. Hilda's Mt. 9 K12 30
St. Hilda's Pl. 9 K12 30
St. Hilda's Rd. 9 K12 30
St. Ives Gro. 12 E11 28
St. Ives Mt. 12 E11 28
St. James App. 14 O9 24
St. James' Av. 18 C6 11
St. James' Dr. 18 C6 11
· St. James' Ter. 18 C6 11
St. James' Wk. 18 C6 11
St. John's Av. 14 Q5 17
St. John's Av. 6 G10 29
St. John's Rd. 3 G10 29
St. Lawrence Cl.
28 A11 26
St. Lawrence Ter.
28 A11 26
St. Luke's Av. 11
Normanton St. H13 37
St. Luke's Cres. 11
Normanton St. H13 37
St. Luke's Gro. 11
Normanton St. H13 37
St. Luke's Mt. 11
Normanton St. H13 37
St. Luke's Pl. 11
Normanton St. H13 37
St. Luke's Rd. 11 H13 37
St. Luke's St. 7 J10 30
St. Luke's St. 11
Normanton St. H13 37
St. Margaret's Av. 8 K8 22
St. Margaret's Av.
18 B6 10
St. Margaret's Cl.
18 B6 10
St. Margaret's Dr.
18 B6 10
St. Margaret's Dr. 8 L8 23
St. Margaret's Gro. 8 L8 23
St. Margaret's Rd.
18 B6 10
St. Margaret's Vw. 8 L8 23
St. Mark's Av. 2 H10 29
St. Mark's Rd. 2 H10 29
St. Mark's Rd. 6 H9 21
St. Mark's St. 2 H10 29
St. Martin's Av. 7 J8 22
St. Martin's Cres. 7 J8 22
St. Martin's Dr. 7 J8 22
St. Martin's Gdns. 7 J8 22

St. Martin's Gro. 7 J8 22
St. Martin's Rd. 7 J8 22
St. Martin's Vw. 7 J8 22
St. Mary's La. 9 K11 30
St. Mary's Rd. 7 J8 22
St. Mary's St. 9 J11 30
St. Matthews Sq. 7
Town St. J7 22
St. Matthew's St. 11 H12 29
St. Matthews Wk. 7 J7 22
St. Matthias' Pl. 4 F10 28
St. Matthias' St. 4 F10 28
St. Michael's Cres. 6 F9 20
St. Michael's Gro. 6 F9 20
St. Michael's La.
4 and 6 F9 20
St. Michael's Rd. 6 F9 20
St. Michael's Ter. 6 F9 20
St. Paul's St. 1 H11 29
St. Peter's Mt. 13 C9 19
St. Peter's Pl. 9 J11 30
St. Peter's St. 9 J11 30
St. Philips Av. 10 H16 42
St. Philip St. 1 H11 29
St. Stephens Pl. 5
Beecroft St. E9 20
St. Stephen's Rd. 9 K11 30
Saint St. 2 J10 30
St. Thomas' Rd. 2 J10 30
St. Thomas St. 28 A10 26
St. Vincent Rd. 28 A12 26
St. Wilfrid's Av. 8 L9 23
St. Wilfrid's Circus, 8 L9 23
St. Wilfrid's Cres. 8 L9 23
St. Wilfrid's Dr. 8 L9 23
St. Wilfrid's Garth, 8 L9 23
St. Wilfrid's Gro. 8 L9 23
Salem Pl. 10 J12 30
Sale St. 3 G11 29
Salisbury Av. 12 F11 28
Salisbury Ct. 18 C6 11
Salisbury Gro. 12 F11 28
Salisbury Mews, 18 C6 11
Salisbury Rd. 12 F11 28
Salisbury Ter. 12
Armley Lodge Rd. F10 28
Salisbury Vw. 12
Armley Lodge Rd. F10 28
Salisbury Vw. 18 C6 11
Salmon Cres. 18 C6 11
Sandbed Cl. 15 P9 25
Sandbed Ct. 15 P9 25
Sandbed La. 15 P9 25
Sandfield Av. 6 · G8 21
Sandford Mt. 5
Beecroft St. E9 20
Sandford Rd. 5 E9 20
Sandhill Cres. 17 K5 14
Sandhill Dr. 17 J5 14
Sandhill Gro. 17 K4 14
Sand Hill La. 17 J5 14
Sandhill Mt. 17 · J4 14
Sandhill Oval, 17 K4 14
Sandhurst Av. 8 L9 23
Sandhurst Gro. 8 L9 23
Sandhurst Pl. 8 L9 23
Sandhurst Rd. 8 L9 23
Sandhurst Ter. 8 L9 23
Sandiford Cl. 15 P9 25 ·
Sandiford Ter. 15
Sandbed La. P9 25
Sandmoor Av. 17 J4 14
Sandmoor Dr. 17 J4 14
Sandon Av. 10
Woodhouse Hill Rd. K14 38
Sandon Gro. 10 K14 38
Sandon Mt. 10
Woodhouse Hill Rd. K14 38
Sandon Pl. 10 K14 38
Sandon St. 10
Woodhouse Hill Rd. K14 38
Sandon Ter. 10
Woodhouse Hill Rd. K14 38
Sandon Vw. 10
Woodhouse Hill Rd. K14 38
Sandringham App. 17 K5 14
Sandringham Cres. 17 J5 14
Sandringham Dr. 17 J5 14

Smithy La. 6
Weetwood La.   F6 12
Smithy La. 16   D4 11
Smools La. 27   E15 36
Snake La. 9   L12 31
Snowden Cl. 13   C9 19
Snowden Cres. 13   C10 27
Snowden Gra. 13   C10 27
Snowden Gro. 13   C10 27
Snowden Royd. 13   C9 19
Snowdon App. 13   C9 19
Society St. 10
Low Rd.   K13 38
Somerdale Cl. 13   C10 27
Somerdale Gdns. 13   C10 27
Somerdale Gro. 13   C10 27
Somerdale Wk. 13   C10 27
Somerset Rd. 28   A11 26
Somers St. 1   H11 29
Somerville Av. 14   N10 32
Somerville Dr. 14   N9 24
Somerville Grn. 14   N10 32
Somerville Mt. 14   O10 32
Somerville Vw. 14   N10 32
South Accommodation
Rd. 10 and 9   J12 30
South Broadgate La.
18   C6 11
South Brook St. 10   J12 30
Southcroft Dr. 10   J16 43
Southcroft Gdns. 10   J16 43
Southcroft Gate, 10   J16 43
Southcroft Grn. 10   J16 43
Southcroft Way, 10   J16 43
South End Av. 13   C10 27
South End Gro. 13   C10 27
South End Mt. 13   C10 27
South End Ter. 13   C10 27
South Farm Cres. 9   M10 31
South Farm Rd. 9   M10 31
South Field Av. 17   K6 14
South Field Dr. 17   K6 14
Southfield Mt. 10
Woodhouse Hill Rd.   K14 38
Southfield Mt. 12
Wesley Rd.   F11 28
Southfield Pl. 12
Strawberry La.   F11 28
Southlands Av. 17   J7 22
Southlands Cres. 17   J7 22
Southlands Dr. 17   J7 22
South Lee, 18   B6 10
Southleigh Av. 11   H15 37
Southleigh Cres. 11   H15 37
Southleigh Dr. 11   H15 37
Southleigh Gdns. 11   H15 37
Southleigh Gro. 11   H15 37
Southleigh Rd. 11   H15 37
Southleigh Vw. 11   H15 37
South Mead Bra.   D2 5
South Par. 28   A12 26
South Par. 1   H11 29
South Par. 6   F8 20
South Pk. Ter. 28   A13 34
South Parkway, 14   N9 24
South Parkway
App. 9   N9 24
South Row, 18   C6 11
South Row, 11   J12 30
Southroyd Par. 28   A12 26
Southroyd Pk. 28   A12 26
Southroyd Rise, 28   A12 26
South Ter. 10
Hunslet La.   J12 30
South Vw. 15
Church La.   P10 33
South Vw. 26   M16 39
South View Rd. 10   K14 38
Southwaite Cl. 14
Brooklands Av.   N9 24
Southwaite Garth, 14
Brooklands Vw.   N9 24
Southwaite La. 14   N9 24
Southwaite Pl. 14
Brooklands Dr.   N9 24
Southway, 18   B5 10
Southwood Cres. 14   P9 25
Southwood Gate, 14   P9 25

Southwood Rd. 14   P9 25
Sovereign St. 1   H11 29
Sowood St. 4   F10 28
Spalding Towers, 9   K10 30
Spark St. 3
Kirkstall Rd.   F10 28
Spark St. 3   G11 29
Speedwell Av. 6
Speedwell St.   H9 21
Speedwell Gro. 6
Speedwell St.   H9 21
Speedwell Mt. 6   H9 21
Speedwell Pl. 6
Speedwell St.   H9 21
Speedwell Rd. 6
Melville Rd.   H9 21
Speedwell St. 6   H9 21
Spen App. 16   D7 19
Spen Bank, 16   D7 19
Spence La. 12   G12 29
Spencer Mt. 8   K9 22
Spencer Pl. 7   K9 22
Spen Cres. 16   D7 19
Spen Dr. 16   E7 20
Spen Gdns. 16   E7 20
Spen Grn. 16   D7 19
Spen La. 5, 6
and 16   E8 to E6 20
Spen Lawn, 16   D7 19
Spennithorne Av. 16   E6 12
Spennithorne Dr. 16   E6 12
Spen Rd. 16   E7 20
Spen Wk. 16   D7 19
Spibey Cres. 18   M15 39
Spibey La. 18   M15 39
Spread Eagle Yd. 11
Meadow La.   J12 30
Spring Av. 27   C15 35
Springbank Av. 28   A9 18
Spring Bank Cres. 6   G9 21
Springbank Dr. 28   A9 18
Springbank Gro. 28   A9 18
Springbank Rise, 28   A9 18
Springbank Rd. 27   C15 35
Springbank Rd. 28   A9 18
Spring Close Row, 9   K12 30
Springfield Av. 10   K14 38
Springfield Cl. 18   D6 11
Springfield Cres. 10
Woodhouse Hill Rd.   K14 38
Springfield Gdns.
18   C6 11
Springfield Gdns.
18   A12 26
Springfield La. BD 4   A14 34
Springfield Mt. 18   C6 11
Springfield Mt. 12   E11 28
Springfield Pl. 10   K14 38
Springfield Pl. 5
Kirkstall La.   E9 20
Springfield Pl.
Pontefract Rd.   L14 39
Springfield Rise, 18   C6 11
Springfield Ter. 5
Kirkstall La.   E9 20
Springfield Ter. 9
Springfield Av.   K14 38
Springfield Ter. 28
Sunfield Pi.   A10 26
Springfield Vw. 5
Kirkstall La.   E9 20
Springfield Wk. 18   C6 11
Spring Gdns. BD 11   A16 34
Spring Gro. 6
Hyde Park Rd.   G10 29
Spring Gro. Av. 6
Hyde Park Rd.   G10 29
Spring Grove Mt. 10   K13 38
Spring Grove Pl. 10   K13 38
Spring Grove St. 10   K13 38
Spring Grove Ter. 6   G10 29
Spring Grove Vw. 10   K13 38
Spring Grove Wk. 6
Alexandra Rd.   G10 29
Springhead Rd. 26   O16 40
Spring Hill, 7   J9 22
Spring Rd. 6   F9 20
Spring St. 13   C9 19

Spring Valley, 28   A10 26
Spring Valley Av. 13   C10 27
Spring Valley Cl. 13   C10 27
Spring Valley
Cres. 13   C10 27
Spring Valley
Croft, 13   C10 27
Spring Valley Dr. 13   C10 27
Spring Valley Mt. 13   C10 27
Spring Valley Vw. 13   C10 27
Springwell Pl. 12   G12 29
Springwell Rd. 12   G12 29
Springwell St. 12   G12 29
Springwell Vw. 11
Holbeck La.   G12 29
Springwood Gro. 8   M8 23
Springwood Rd. 8   L8 23
Spink Rd. 9   K10 30
Stainbeck Av. 7   G8 21
Stainbeck Cres. 7   G8 21
Stainbeck Gdns. 7   H8 21
Stainbeck Gro. 7   G8 21
Stainbeck La.
6 and 7   G7 to J7 21
Stainbeck Rd. 7   H7 21
Stainbeck Side, 7   G8 21
Stainbeck Sq. 7   G8 21
Stainbeck St. 7   G8 21
Stainbeck Ter. 7   G8 21
Stainbeck Vw. 7   G8 21
Stainbeck Wk. 7   H8 21
Stainburn Av. 17   K6 14
Stainburn Cres. N. 17   J6 14
Stainburn Cres. S. 17   J6 14
Stainburn Dr. 17   J6 14
Stainburn Mt. 17   K7 22
Stainburn Rd. 17   J7 22
Stainburn Ter. 17   J7 22
Stainburn Vw. 17   K6 14
Staincross Rd. 11
Croydon Pl.   G12 29
Staincross Ter. 11
Croydon Pl.   G12 29
Stainmore Cl. 14   O9 24
Stairfoot La.
16 and 17   G4 13
Staithe Av. 10   J16 43
Staithe Cl. 10   J16 43
Staithe Gdns. 10   J16 43
Stafford Chase, 10   J13 38
Stafford St. 10   K13 38
Stamford Pl. 10
Skinner La.   J10 30
Stamford St. 7   J10 30
Standard Pl. 10
Norwich Av.   J11 38
Standard St. 10
Norwich Av.   J14 38
Stanhope Av. 18   C6 11
Stanhope Dr. 18   B7 18
Stanhope Pl. 7   J9 22
Stanhope Sq. 7
Sheepscar St.   J9 22
Stanhope Ter. 7   J9 22
Stanks App. 14   P9 25
Stanks Av. 14   P9 25
Stanks Cl. 14   Q9 25
Stanks Cross, 14   P9 25
Stanks Dr. 14   Q9 25
Stanks Gdns. 14   P8 25
Stanks Grn. 14   Q9 25
Stanks La. 14   P9 25
Stanks La. N. 14   P8 25
Stanks La. S. 14   P9 25
Stanks Par. 14   P9 25
Stanks Rise, 14   P8 25
Stanks Rd. 14   P9 25
Stanley Av. 9
Ashley Rd.   K10 30
Stanley Dr. 8   L6 15
Stanley Pl. 9
Compton Av.   L10 31
Stanley Rd. 9   K10 30
Stanley Ter. 12
Colton Rd.   F11 28
Stanley Vw. 12   E11 28
Stanmore Av. 4   F9 20

74

Telford Wk. 10 K14 38
Temperance St. 5
  Club Row E9 20
Tempest Cres. 11 H13 37
Tempest Rd. 11 H13 37
Templar La. 2 J11 30
Templar St. 2 J11 30
Templar Villas, 15 P9 25
Temple Av. 26 N15 40
Temple Av. 15 O12 32
Temple Cl. 15 O12 32
Temple Ct. 26 N15 40
Temple Ct. 15 N11 32
Temple Cres. 11
  Cemetery Rd. H13 37
Temple Gate, 15 O11 32
Temple Gate Av. 15 O12 32
Temple Gate Cl. 15 O11 32
Temple Gate Cres. 15 O12 32
Temple Gate Dr. 15 O11 32
Temple Gate Gro. 15 O12 32
Temple Gate Rise, 15 O12 32
Temple Gate Rd. 15 O12 32
Temple Gate Vw. 15 O12 32
Temple Gate Wk. 15 O11 32
Temple Gate Way, 15 O12 32
Temple Grn. 15 O11 32
Temple Gro. 15 O11 32
Temple La. 15 O11 32
Temple Lea, 15 O11 32
Templenewsam Rd. 15 N11 32
Templenewsam Vw. 15 N12 32
Temple Park Cl. 15 O11 32
Temple Park Gdns. 15 O11 32
Temple Park Grn. 15 O11 32
Temple Rise, 15 O12 32
Temple Stowe Cres. 15 P10 33
Temple Stowe Dr. 15 P11 33
Temple Stowe Gdns. 15 O11 32
Temple Stowe Hill, 15 O10 32
Temple Vw. 9 K12 30
Temple View Pl. 9 K11 30
Temple View Rd. 9 K11 30
Temple View Ter. 9 K11 30
Temple Wk. 15 O11 32
Tennyson St. 28 A12 26
Tenter La. 1
  Bridge End J11 30
Tenth Av. 12
  Eighth Av. G12 29
Theaker La. 12 E11 28
Theaker Vw. 12
  Theaker La. E11 28
Thealby Cl. 9 K11 30
Thealby Lawn, 9 K10 30
Thealby Pl. 9 K10 30
Theodore St. 11 H14 37
Third Av. 26 N15 40
Third Av. 12 G11 29
Thirlmere Gdns. 11 G15 37
Thirteenth Av. 12
  Ninth Av. F12 28
Thomas St. 6 ·H9 21
Thomas Yd. 1
  Cankerwell La. H10 29
Thoresby Pl. 1 H11 29
Thorn Cl. 8 L9 23
Thorn Cres. 8 L9 23
Thorn Dr. 8 L9 23
Thornfield Rd. 16 E7 20
Thorn Gro. 8 L9 23
Thornhill Pl. 12 E12 28
Thornhill Rd. 12 E12 28
Thornhill St. 12 E12 28
Thornhill Ter. 12 E11 28

Thornhill Vw. 12 E12 28
Thorn La. 8 K7 22
Thornleigh Mt. 9 K12 30
Thornleigh St. 9
  Easy Rd. K12 30
Thorn Mt. 8 M9 23
Thorn Ter. 8 L9 23
Thornton Av. 12 E11 28
Thornton Gdns. 12 E11 28
Thornton Gro. 12 E11 28
Thornton's Arc. 1
  Briggate J11 30
Thorn Vw. 8 L9 23
Thornville Av. 6 G10 29
Thornville Cres. 6
  Brudenell Rd. G9 21
Thornville Gro. 6 G10 29
Thornville Mt. 6
  Harold Av. G10 29
Thornville Pl. 6
  Thornville St. G10 29
Thornville Rd. 6 G10 29
Thornville Row, 6
  Thornville St. G10 29
Thornville St. 6 G10 29
Thornville Vw. 6
  Thornville St. G10 29
Thorn Wk. 8 M9 23
Thorpe Av. Thorpe J17 43
Thorpe Cres. Thorpe J18 43
Thorpe Cres. 10 J17 43
Thorpe Gdns. 10 J17 43
Thorpe Gro. 10 J17 43
Thorpe La. 10 and Thorpe H18 43
Thorpe Lower La. Thorpe K18 43
Thorpe Mt. 10 H17 42
Thorpe Pl. 10 J17 43
Thorpe Rd. 28 A11 26
Thorpe Rd. 10 J17 43
Thorpe Sq. 10 J17 43
Thorpe St. 15 O11 32
Thorpe St. 10 J17 43
Thorpe Ter. 10 J17 43
Thorpe Vw. 10 J17 43
Throstle Av. 10 H17 42
Throstle Dr. 10 H17 42
Throstle Gro. 10 J17 43
Throstle Hill, 10 H17 42
Throstle La. 10 H17 42
Throstle Mt. 10 H17 42
Throstle Nest Vw. 18 C7 19
Throstle Par. 10 H17 42
Throstle Pl. 10 H17 42
Throstle Rd. 10 J17 to K16 43
Throstle Row, 10 H17 42
Throstle Sq. 10 J17 43
Throstle St. 10 H17 42
Throstle Ter. 10 J17 43
Throstle Vw. 10 J17 43
Throstle Wk. 10 H17 42
Throston Rd. 8 M9 23
Thwaite La. 10 L13 39
Tilbury Av. 11
  Tilbury Rd. G13 37
Tilbury Gro. 11
  Tilbury Rd. G13 37
Tilbury Mt. 11 G13 37
Tilbury Par. 11
  Tilbury Rd. G13 37
Tilbury Pl. 11 G13 37
Tilbury Rd. 11 G13 37
Tilbury Ter. 11 G13 37
Tilbury Vw. 11 G13 37
Tile La. 16 F5 12
Timber Pl. 9 K12 30
Tinshill Av. 16 D5 11
Tinshill Cl. 16 D5 11
Tinshill Cres. 16 D5 11
Tinshill Dr. 16 D4 11
Tinshill Garth, 16 D5 11
Tinshill Gro. 16 D5 11
Tinshill La. 16 C5 11
Tinshill Mt. 16 D5 11
Tinshill Rd. 16 C5 11

Tinshill Vw. 16 D5 11
Toft Pl. 12
  Toft St. F12 28
Tofts House Cl. 28 A11 26
Tofts Rd.28 A11 26
Toft St. 12 F12 28
Tong App. 12 C12 27
Tong Dr. 12 C11 27
Tong Gate, 12 C11 27
Tong Grn. 12 C11 27
Tong Rd. BD 11
  and 12 A14 to G12 34
Tongue La. 6 G7 21
Tong Wk. 12 C11 27
Tong Way, 12 C11 27
Top Fold, 12 E12 28
Topham St. 12 F12 28
Top Moor Side, 11 H12 29
Torbay Av. 11
  Torbay St. G12 29
Torbay Gro. 11
  Torbay St. G12 29
Torbay Pl. 11
  Torbay St. G12 29
Torbay St. 11 G12 29
Tordoff Ter. 5 E9 20
Toronto Pl. 7 J8 22
Toronto St. 1 H11 29
Torre Cl. 9 L11 31
Torre Cres. 9 L11 31
Torre Dr. 9 L10 31
Torre Gdns. 9 K11 30
Torre Grn. 9 K11 30
Torre Gro. 9 L10 31
Torre Hill, 9 L10 31
Torre La. 9 L11 31
Torre Mt. 9 L10 31
Torre Rd. 9 K11 30
Torre Sq. 9 L10 31
Torre Vw. 9 L10 31
Torre Wk. 9 L10 31
Tower Gro. 12 E11 28
Tower La. 12 E11 28
Tower Pl. 12 D11 27
Towers Way, 6 H7 21
Town End, 27 C15 35
Town Gate, 11 H12 29
Town St. 27 C15 35
Town St. 18 B7 18
Town St. 19 A5 10
Town St. 13 A8 18
Town St. 28 A10 26
Town St. 17 M2 9
Town St. 10 H16 to K16 42
Town St. 11 G14 37
Town St. 12 C11 20
Trafalgar Row, 11 J12 30
Trafalgar St. 2 J11 30
Trafalgar Ter. 7
  Albert Gro. F8 20
Trafford Av. 9 L10 31
Trafford Gro. 9 L9 23
Trafford Ter. 9
  Seaforth Av. L9 23
Tranquility, 15
  Tranquility Av. P10 33
Tranquility Av. 15 P10 33
Tranter Pl. 15 N11 32
Tredgold Av. 16 D2 5
Tredgold Garth, 16 D2 5
Trelawn Av. 6 F8 20.
Trelawn Cres. 6 F8 20
Trelawn Pl. 6 F8 20
Trelawn St. 6 F8 20
Trelawn Ter. 6 F8 20
Tremont Ter. 10
  Woodhouse Hill Rd. K14 38
Tremont Vw. 10
  Woodhouse Hill Rd. K14 38
Trenic Cres. 6 F9 20
Trenic Dr. 6 F9 20
Trentham Av. 11
  Stratford Ter. H13 37
Trentham Gro. 11
  Stratford Ter. H13 37
Trentham Pl. 11
  Stratford Ter. H13 37

Trentham Row, 11
Stratford Ter.   H13 37
Trentham St. 11   H14 37
Trent Rd. 9   L10 31
Tresco Av. 13   D10 27
Trilby St. 13   B10 26
Trinity St. 1   J11 30
Troughton Pl. 28   A12 26
Troughton St. 28   A12 26
Troydale Gdns. 28   B12 26
Troydale Gro. 28   B12 26
Troydale La. 28
and 12   B12 26
Troy Hill, 18   C6 11
Troy Rd. 18   C6 11
Tudor Gdns. 11   G14 37
Tunstall Rd. 11   J13 38
Turkey Hill, 28   A12 26
Turk's Head Yd. 1
Briggate   J11 30
Turnaways, The, 6   F9 20
Twelfth Av. 12
Eighth Av.   G12 29
Tyas Gro. 9   L11 31
Tynwald Cl. 17   H6 13
Tynwald Dr. 17   H5 13
Tynwald Hill, 17   H6 13
Tynwald Mt. 17   H6 13
Tynwald Rd. 17   H6 13
Tynwald Wk. 17   H5 13

Ullswater Cres. 15   N11 32
Una Mt. 11
Folly La.   H13 37
Union Gro. 7   J8 22
Union Pl. 11   H12 29
Union Rd. 12
Parliament Rd.   F11 28
Union St. 2   J11 30
Union Ter. 7   J8 22
Upland Cres. 8   L8 23
Upland Gro. 8   L8 23
Upland Rd. 8   L9 23
Upper Accommodation
Rd. 9   K11 30
Upper Burmantofts St. 9
Burmantofts St.   K11 30
Upper Carr Pl. 10   J14 38
Upper Cobden Pl. 2
Cobden Pl.   D13 35
Upper Fountaine St.
2   H11 29
Upper Mill Hill, 1
Boar La.   H11 29
Upper Tower La. 12   D11 27
Upper Town St. 13   C9 19
Upper Westlock Av.
9   L10 31
Upper Woodview Pl. 11
Woodview St.   H14 37
Upper Wortley Rd.
12   E11 28
Upper Wortley Vw. 12
Barras Garth Pl.   E12 28
Ursula St. 11
Malvern St.   H13 37

Vale Av. 8   L6 15
Vale, The, 6   G8 21
Valley Cl. 17   H4 13
Valley Dr. 15   O10 32
Valley Gdns. 7   J7 22
Valley Gro. 15   O10 32
Valley Mt. 13   B11 26
Valley Rise, 13   C8 19
Valley Rd. 28   A12 26
Valley Rd. 13   C8 19
Valley Sq. 28   A12 26
Valley Ter. 17   K5 14
Valley The, 17   H4 13
Vancouver Pl. 7
Montreal Av.   J8 22
Varley St. 28   A10 26
Vaux St. 10   J12 30
Venetian Pl. 6
Institution St.   H9 21

Ventnor St. 3   G10 29
Verity Spur, 9   N11 32
Verity Vw. 9   N11 32
Vermont St. 13   B10 26
Vernon Pl. 28   A10 26
Vernon St. 2   H10 29
Vesper Gdns. 5   D8 19
Vesper Gate Cres. 5   D8 19
Vesper Gate Dr. 5   D8 19
Vesper Gate Mt. 5   D8 19
Vesper Gro. 5
Norman St.   E9 20
Vesper La. 5   D8 19
Vesper Mt. 5
Norman St.   E9 20
Vesper Pl. 5
Norman St.   E9 20
Vesper Rise, 5   D8 19
Vesper Rd. 5   D8 19
Vesper Ter. 5
Norman St.   E9 20
Vesper Wk. 5   D8 19
Vesper Way, 5   D8 19
Vevers Ter. 11
Nineveh Rd.   H12 29
Viaduct Rd. 4   F10 28
Vicarage Av. 5   E9 20
Vicarage Pl. 5   E9 20
Vicarage Rd. 6   G10 29
Vicarage Rd. 15   P9 25
Vicarage St. 5   E9 20
Vicarage Ter. 5   E9 20
Vicarage Vw. 5   E9 20
Vicar La. 1   J11 30
Vicar's Rd. 8
Shepherd's La.   K9 22
Vicar's Ter. 8
Shepherd's La.   K9 22
Vicar St. 11   J13 38
Vickers Av. 5   D9 19
Vickers Dale, 28
Town St.   A10 26
Vickers Pl. 28   A10 26
Vickers St. 28   A10 26
Vickers Ter. 28   A10 26
Victoria Av. 9   L11 31
Victoria Cl. 18   B7 18
Victoria Cres. 18   B7 18
Victoria Dr. 18   B7 18
Victoria Gdns. 18   B7 18
Victoria Gro. 7
Carr La.   J11 30
Victoria Gro· 9   L11 31
Victoria Mt. 18   B7 18
Victoria Park Av.
13 and 5   D9 19
Victoria Park Gro. 5   D9 19
Victoria Rd. 18   B6 10
Victoria Rd. 27   E16 36
Victoria Rd. 5   E9 20
Victoria Rd. 6   G9 21
Victoria Rd. 11   H12 29
Victoria Row, 28 & 13
Swinnow Rd.   A11 26
Victoria Sq. 1   H11 29
Victoria St. 3   G10 29
Victoria St. 7   J7 22
Victoria St. 10   J12 30
Victoria Ter. 3   G10 29
Victoria Wk. 18   B7 18
View, The, 8   K7 22
View, The, 17   G4 13
Village Pl. 4   F10 28
Village Rd. 16   G2 7
Village Ter. 4   F9 20
Village, The, St. 4   F10 28
Villa Ter. 11
Woodhouse Hill Rd. K14 38
Vincent Pl. 10   J12 30
Vinery Av. 9   L11 31
Vinery Gro. 9
Vinery Av.   L11 31
Vinery Mt. 9   L11 31
Vinery Pl. 9   L11 31
Vinery Rd. 4   F10 28
Vinery Ter. 9   L11 31
Vinery Vw. 9   L11 31
Vine St. 10   J12 30

Virginia Ter. 2
Mount Preston   H10 29
Vulcan St. 7   J10 30

Wade La. 2   J11 30
Wade St. 2   J11 30
Wade Yd. 2   J11 30
Waincliffe Cres. 11   G14 37
Waincliffe Dr. 11   G14 37
Waincliffe Garth, 11
Waincliffe Pl.   G14 37
Waincliffe Mt. 11   G14 37
Waincliffe Pl. 11   G14 37
Waincliffe Sq. 11   G14 37
Waincliffe Ter. 11   G15 37
Wainright Ter. 10
Church St.   J13 38
Wakefield Av. 14   N10 32
Wakefield Rd. 26
   Q15 to S12 41
Wakefield Rd. 10 and
Roth.   L14 to L16 39
Walford Av. 9
Nickleby Rd.   L11 31
Walford Gro. 9
Nickleby Rd.   L11 31
Walford Mt. 9
Nickleby Rd.   L11 31
Walford Rd. 9   L11 31
Walford Ter. 9
Nickleby Rd.   L11 31
Walker Rd. 18   B6 10
Walker's Ct. 10
Church St.   J13 38
Walker's La. 12   F13 36
Walkers Rd. 6   G8 21
Walmsley Rd. 6   G9 21
Walsh La. 12   C14 35
Walter Cres. 9   K11 30.
Walter Gro. 7
Oxford Rd.   H9 21
Walter Vw. 9   K11 30
Walworth Pl. 6
Alexandra Rd.   G10 29
Ward St. 10   J12 30
Warehouse Hill, 2
Call La.   J11 30
Warrel's Av. 13   C9 19
Warrel's Gro. 13   C10 27
Warrel's Mt. 13   C10 27
Warrel's Pl. 13   C9 19
Warrel's Rd. 13   C9 19
Warrel's St. 13   C10 27
Warrel's Ter. 13
Granhamthorpe   C10 27
Warren Pl. 6
Institution St.   H9 21
Warwickshire St. 10   K13 38
Warwick Ter. 1
Caledonian Rd.   H10 29
Washington Pl. 13
Harley Gdns.   B11 26
Washington St. 3   G11 29
Water Ct. 10
Water La.   H12 29
Water Hall, 11
Water La.   H12 29
Water La. 18   A6 10
Water La. 11   H12 29
Water La. 12   C11 27
Waterloo Cres. 13   C9 19
Waterloo La. 13   C9 19
Waterloo Rd. 10   K13 38
Waterloo St. 10   J11 30
Waterloo Vw. 11
Elland Rd.   H12 29
Water Mt. 13   C9 19
Waterton Pl. 10
Grove Rd.   K13 38
Waterton Rd. 10
Grove Rd.   K13 38
Watlass St. 4   F10 28
Watson Mt. 5
Commercial Rd.   E9 20
Watson Rd. 14   N10 32
Watson's Yd. 6
St. Marks Rd.   H9 21

Waveney Rd. 12   F12 28
Waver Grn. 28   A11 26
Waverley Gro. 11
  Lady Pit St.   H13 37
Waverley Mt. 11
  Lady Pit St.   H13 37
Waverley Pl. 11
  Lady Pit St.   H13 37
Waverley Ter. 7   K9 22
Wayland Dr. 16   F5 12
Wayland Rise, 16   F5 12
Wayland Wk. 16   F5 12
Weavers Sq. 9
  Richmond St.   K13 38
Weaver St. 4   F10 28
Webster Row, 12   E12 28
Websters Fold, 11
  Town St.   G14 37
Wedgewood Dr. 8   L7 23
Wedgewood Gro. 8   L7 23
Weetwood Av. 16   F7 20
Weetwood Ct. 16   F7 20
Weetwood Cres. 16   F7 20
Weetwood Grange
  Gro. 16   E7 20
Weetwood Grange Rd.
  16   E7 20
Weetwood La. 16   F6 12
Weetwood Park Dr.
  16   E7 20
Weetwood Rd. 16   E7 20
Weetwood Ter. 16   F7 20
Weetwood Wk. 16   F7 20
Welbeck Rd. 9   L11 31
Welburn Av. 16   E7 20
Welburn Dr. 16   E7 20
Welburn Gro. 16   E7 20
Weldon Pl. 11   H12 29
Well Close Mt. 7   H10 29
Well Close Pl. 2   H10 29
Well Close Rd. 7   H10 29
Well Close Ter. 7   J10 30
Well Close Vw. 7   J10 30
Weller Av. 9   K10 30
Weller Gro. 9   K10 30
Weller Mt. 9   K10 30
Weller Pl. 9   K10 30
Weller Rd. 9   K10 30
Weller Ter. 9   K10 30
Weller Vw. 9   K10 30
Wellfield Pl. 6
  North La.   F9 20
Well Garth, 15   P10 33
Well Garth Bank, 13   B9 18
Well Garth Mt. 15   O10 32
Well Garth Pl. 12
  Armley Rd.   F11 28
Well House Av. 8   L8 23
Well House Cres. 8   L8 23
Well House Dr. 8   L8 23
Well House Gdns. 8   L8 23
Well House Rd. 8   L8 23
Wellhouse Sq. 7   J8 22
Wellington Bridge Rd.
  3   G11 29
Wellington Garth,
  13   C9 19
Wellington Gro. 13   C9 19
Wellington La. 1   H11 29
Wellington Mt. 13   C9 19
Wellington Rd. 12   G12 29
Wellington St. 1   G11 29
Wellington Ter. 13   C9 19
Well Lane, 7   J7 22
Well Pit Cl. 12
  School Cl.   C13 35
Wells Croft, 6   G7 21
Wellstone Av. 13   B10 26
Wellstone Dr. 13   B10 26
Wellstone Gdns. 13   B11 26
Wellstone Garth, 13   B11 26
Wellstone Grn. 13   B10 26
Wellstone Rise, 13   B11 26
Wellstone Rd. 13   B11 26
Wellstone Way, 13   B11 26
Welton Gro. 6   G9 21
Welton Mt. 6   G9 21
Welton Pl. 6   G9 21

Welton Rd. 6   G9 21
Wensley Av. 7   J7 22
Wensley Cres. 7   J7 22
Wensley Dr. 7   H7 21
Wensley Gdns. 7   H7 21
Wensley Grn. 7   H7 21
Wensley Gro. 7   H7 21
Wensley Rd. 7   H7 21
Wensley Vw. 7   J7 22
Wepener Mt. 9   L10 31
Wepener Pl. 9   L10 31
Wesley Av. 12
  Athlone St.   F11 28
Wesley Cl. 11   G13 37
Wesley Ct. 11   G14 37
Wesley Croft, 11   G13 37
Wesley Garth, 11   G13 37
Wesley Pl. 9   K11 30
Wesley Rd. 12   F11 28
Wesley St. 28   A9 18
Wesley St. 2   J10 30
Wesley St. 11   G13 37
Wesley St. 13
  Town St.   A8 18
Wesley Ter. 13   A8 18
Wesley Ter. 13   C9 19
Wesley Vw. 13   A8 18
West Av. 8   M7 15
Westbourne Av. 11
  Rowland Rd.   H13 37
Westbourne Mt. 11
  Rowland Rd.   H13 37
Westbourne Pl. 11
  Rowland Rd.   H13 37
Westbourne Pl. 28
  Richardshaw La.   A10 26
Westbourne St. 11
  Rowland Rd.   H13 37
Westbrook Cl. 18   B6 10
Westbrook La. 18   B6 10
Westbury Gro. 10
  Parnaby Rd.   K14 38
Westbury Mt. 10   K14 38
Westbury Pl. 10   K14 38
Westbury St. 10   K14 38
Westbury Ter. 10
  Parnaby Rd.   K14 38
Westcombe Av. 8   L6 15
West End, 12
  Forge Row   C13 35
West End Cl. 18   A6 10
West End Dr. 18   A6 10
West End Gro. 18   A6 10
West End La. 18   A6 10
West End Rise, 18   A6 10
West End Ter. 6   G9 21
Westerley Croft, 12   F11 28
Western Gro. 12
  Lower Wortley Rd.   E12 28
Western Mt. 12
  Lower Wortley Rd.   E12 28
Western Rd. 12   E12 28
Western St. 12
  Lower Wortley Rd.   E12 28
Western Ter. 12
  Lower Wortley Rd.   E12 28
Western Vw. 12
  Lower Wortley Rd.   E12 28
West Farm Av. 10   H16 42
Westfield, 12   D11 27
Westfield Av. 12   D11 27
Westfield Cres. 3   G10 29
Westfield La. 14   Q5 17
Westfield Pl. 3
  Westfield Rd.   G10 29
Westfield Rd. 3   G10 29
Westfield St. 3
  Westfield Rd.   G10 29
Westfield Ter. 7   J7 22
Westgate, 1   H11 29
West Grange Cl. 10   J14 38
West Grange Dr. 10   J14 38
West Grange Fold,
  10   J14 38
West Grange Gdns.
  10   J14 38
West Grange Grn. 10   J14 38
West Grange Rd. 10   J15 38

West Grange Wk. 10   J14 38
West Grove St. 28
  Richardshaw La.   A10 26
West Hill Av. 7   J7 22
Westland Rd. 11   H14 37
Westland Sq. 11   H15 37
West Lea Cl. 6   H6 13
West Lea Ct. 6   H6 13
West Lea Dr. 6   H6 13
West Lea Gdns. 17   H6 13
West Lea Garth, 6   H6 13
Westlock Av. 9   K10 30
Westlock Cres. 9   K10 30
Westlock Gro. 9   K10 30
Westlock Ter. 9   K10 30
West Lodge Gdns. 7   J8 22
Westminster Cres.
  15   N11 32
Westminster Pl. 1
  West St.   G11 29
Westmoor Pl. 13
  Westmoor St.   B9 18
Westmoor Rise, 13   B9 18
Westmoor Rd. 13   B9 18
Westmoor St. 13   B9 18
Westmorland Mt. 13
  Waterloo La.   C9 19
West Mount St. 11
  South Ridge St.   H13 37
Weston Av. 11
  Weston Gro.   H12 29
Weston Gro. 11   H12 29
Weston Mt. 11
  Weston Gro.   H12 29
Weston Ter. 11   H12 29
Weston Vw. 11   H12 29
Westover Rd. 13   C9 19
Westover St. 13
  Westover Rd.   C9 19
West Parade, 16   E7 20
West Park Av. 8   L5 15
West Park Cl. 8   L5 15
West Park Cres. 8   L6 15
West Park Dr. 16   E7 20
West Park Dr. E. 8   L5 15
West Park Dr. W. 8   K5 14
West Park Gro. 8   L5 15
West Park Pl. 8   L6 15
West Park Rd. 8   L6 15
West St. 28   A11 26
West St. 1   G11 29
West Vw. St. 11   H13 37
West Wood Ct. 10   H16 42
Westwood Gro. 11
  Northcote Rd.   H13 37
West Wood Rd. 10   H16 42
Westwood St. 11
  Northcote Rd.   H13 37
Westwood Ter. 11
  Northcote Rd.   H13 37
Westwood Vw. 11
  Northcote Rd.   H13 37
Wetherby Gro. 4
  Burley Rd.   F10 28
Wetherby Pl. 4
  Burley Rd.   F10 28
Wetherby Rd. 8, 14, 17,
       L8 to P3 23
Wetherby Ter. 4
  Burley Rd.   F10 28
Wharfedale Av. 7   H9 21
Wharfedale Gro. 7
  Wharfedale Av.   H9 21
Wharfedale Mt. 7
  Wharfedale Av.   H9 21
Wharfedale Pl. 7
  Wharfedale Av.   H9 21
Wharfedale St. 7
  Wharfedale Av.   H9 21
Wharfedale Ter. 7   H9 21
Wharfedale Vw. 7
  Wharfedale Av.   H9 21
Wharf St. 2   J11 30
Wheater Pl. 10
  Royal Rd.   J14 38
Wheater St. 10   J14 38
Wheaton Av. 15   O11 32
Wheelwright Av. 12   E12 28

Wheelwright Cl. 12 E12 28
Whinbrook Cres. 17 J6 14
Whinbrook Gdns. 17 J6 14
Whinbrook Gro. 17 J6 14
Whincover Bank, 12 D12 27
Whincover Cl. 12 D12 27
Whincover Cross,
12 D12 27
Whincover Dr. 12 D12 27
Whincover Gdns. 12 D12 27
Whincover Gro. 12 D12 27
Whincover Hill, 12 D12 27
Whincover Mt. 12 D12 27
Whincover Rd. 12 D12 27
Whincover Vw. 12 D12 27
Whinfield, 16 E5 12
Whingate, 12 E11 28
Whingate Av. 12 E11 28
Whingate Cl. 12 E11 28
Whingate Gro. 12 E11 28
Whingate Rd. 12 E11 28
Whingate Ter. 12 E11 28
Whinmoor Way, 14 P8 25
Whinmore Ct. 14 J16 43
Whinmore Cres. 14 O6 16
Whinmore Gdns. 14 N6 16
Whitebridge Av. 9 N11 32
Whitebridge Cres. 9 N10 32
Whitebridge Spur, 9 N10 32
Whitebridge Vw. 9 N10 32
Whitechapel Yd. 11
Meadow La. J12 30
Whitecote Gdns. 13 B9 18
Whitecote Hill, 13 B9 18
Whitecote Lane, 13 B8 18
Whitecote Mt. 13
Newlay La. C9 19
Whitecote Rise, 13 B9 18
Whitecote Sq. 13 C9 19
Whitecote St. 13
Leeds & Bradford Rd. C9 19
White Gro. 8 L7 23
Whitehall Rd. BD 11 12
and 1 A16 to H11 34
White Horse Ct. 1
Boar La. H11 29
Whitelands Bldgs.
28 A11 26
Whiteley St. 12 F11 28
Whitelock St. 7 J10 30
White Rose Yd. 6
Woodhouse Cliff H9 21
Whitfield St. 8 K9 22
Whitfield St. 10 K13 38
Whitkirk Cl. 15 P11 33
Whitkirk La. 15 P11 33
Wickham St. 11 H13 37
Wide La. 27 G17 42
Wigton La. 17 K4 to L4 14
Wilfred Av. 15 O11 32
Wilfred St. 12 E13 36
Wilfred St. 15 O11 32
Wilfred Ter. 12 E13 36
William Rise, 15 N11 32
William St. 27 E15 36
William St. 28 A10 26
William St. 10 J12 30
William Vw. 15 N11 32
Williamson St. 11
Domestic St. G12 29
Willians La. 26 N15 40
Willis St. 9 K11 30
Willoughby Cres. 11 G12 29
Willoughby Gro. 11 G12 29
Willoughby Mt. 11 G12 29
Willoughby Pl. 11 G12 29
Willoughby Rd. 11 G12 29
Willoughby St. 11 G12 29
Willoughby Ter. 11
Domestic St. G12 29
Willoughby Vw. 11 G12 29
Willow Cres. 15 N11 32
Willow Grove Rd. 1
Tonbridge St. H10 29
Willow Rd. 4 G10 29
Willows, The, 17
Street La. J6 14

Willow Well Rd.
15 N11 32
Wilmington Gro. 7 J9 22
Wilmington St. 7 J10 30
Wilmington Ter. 7 J9 22
Wilson Ct. 12
Wellington Rd. G12 29
Wilsons Pl. 11
Great Wilson St. H12 29
Wilson St. 10 K13 38
Wilson Gro. 6 G8 21
Wilton Pl. 12
Parliament Rd. F11 28
Wilton St. 12
Parliament Rd. F11 28
Wilton Ter. 12 F11 28
Wiltshire Pl. 10 K13 38
Wiltshire St. 10 K13 38
Wiltshire Ter. 10 K13 38
Winchester Mt. 12 F11 28
Winchester Pl. 12 F11 28
Winchester Rd. 12 F11 28
Winchester St. 12 F11 28
Winchester Ter. 12 F11 28
Winchester Vw. 12 F11 28
Winding Way, 17 H4 13
Windmill App. 10 K15 38
Windmill Mt. 10 J15 38
Windmill Rd. 10 J15 38
Windmill Yd. 26 M16 39
Windsor Av. 15 O11 32
Windsor Mt. 15 O11 32
Windsor Pl. 10
New Pepper Rd. K13 38
Windsor Ter. 10
New Pepper Rd. K13 38
Wine St. 1
Infirmary St. H11 29
Winfield Gro. 2
Devon Rd. H10 29
Winfield Mt. 7 H10 29
Winfield Pl. 2 H10 29
Winfield Rd. 7 H10 29
Winfield Ter. 2
Devon Rd. H10 29
Wingham St. 7 J10 30
Winnie Ter. 12 G12 29
Winnipeg Pl. 7
King George Av. J7 22
Winn Moor La. 17 N5 16
Winrose App. 10 K15 38
Winrose Av. 10 J15 38
Winrose Cres. 10 J15 38
Winrose Dr. 10 J15 38
Winrose Garth, 10 J15 38
Winrose Gro. 10 K15 38
Winrose Hill, 10 K14 38
Winstanley Ter. 6 G9 21
Winston Gdns. 6 F8 20
Winston Mt. 6 F8 20
Winthorpe St. 6 G8 21
Wintoun St. 7 J10 30
Woerth Pl. 7
Camp Rd. J10 30
Wolley Av. 12 C13 35
Wolley Dr. 12 C13 35
Wolley Gdns. 12 C13 35
Wolscot St. 11
Ladbroke St. H12 29
Wolseley Av. 4
Burley Rd. F10 28
Wolseley Cres. 4
Burley Rd. F10 28
Wolseley Gro. 4 F10 28
Wolseley Mt. 4
Wolseley Rd. F10 28
Wolseley Pl. 7 J9 22
Wolseley Rd. 4 F10 28
Wolseley St. 4 F10 28
Wolseley Ter. 7 J9 22
Woodbine Ter. 6 G12 29
Woodbine Ter. 13 C9 19
Woodbridge Cl. 6 E8 20
Woodbridge Cres. 6 E8 20
Woodbridge Fold, 6 E8 20
Woodbridge Gdns. 6 E8 20
Woodbridge Garth, 6 E8 20
Woodbridge Grn. 6 E8 20

Woodbridge Lawn, 6 E8 20
Woodbridge Pl. 6 E8 20
Woodbridge Rd. 6 E8 20
Woodbridge Vale, 6 E8 20
Woodbourne Av. 17 H6 13
Woodfield Ter. 28 A12 26
Woodhall Av. 5 D8 19
Woodhall Dr. 5 D8 19
Woodhead La. 27 B15 34
Wood Hill Cres. 16 C5 11
Wood Hill Gdns. 16 C5 11
Wood Hill Garth, 16 C5 11
Wood Hill Gro. 16 C5 11
Wood Hill Rise, 16 C5 11
Wood Hill Rd. 16 C5 11
Woodhouse Cliff, 6 H9 21
Woodhouse Hill Av. 10
Woodhouse Hill Rd. K14 38
Woodhouse Hill Cres. 10
Woodhouse Hill Rd. K14 38
Woodhouse Hill Gro. 10
Woodhouse Hill Rd. K14 38
Woodhouse Hill Pl. 10
Woodhouse Hill Rd.
10 K14 38
Woodhouse Hill St. 10
Woodhouse Hill Rd. K14 38
Woodhouse Hill Ter. 10
Woodhouse Hill Rd. K14 38
Woodhouse Hill Vw. 10
Woodhouse Hill Rd. K14 38
Woodhouse La.
6 and 2 H9 to J11 21
Woodhouse Sq. 2 H10 29
Woodhouse St. 6 H9 21
Woodland Cl. 15 P11 33
Woodland Croft, 18
C6 11
Woodland Dr. 7 J7 22
Woodland Gro. 7 K9 22
Woodland Hill, 15 O11 32
Woodland La. 7 J7 22
Woodland Mt. 7 K9 22
Woodland Park Rd. 6 G8 21
Woodland Rise, 15 P11 33
Woodland Rd. 15 O11 32
Woodland Ter. 26
M16 39
Woodland Ter. 7 H7 21
Woodland Vw. 7 J7 22
Woodlands Ct. 16 E6 12
Woodlands Pk. Gro.
28 A12 26
Woodlands Park Rd.
28 A12 26
Wood La. 18 C7 19
Wood La. 26
L15 to N16 39
Wood La. 6 F8 20
Wood La. 7 J7 22
Wood La. 12 C14 35
Wood La. 12 C11 27
Woodlea Mt. 11 G13 37
Woodlea Pl. 11 G13 37
Woodlea St. 11 G13 37
Woodcliffe Cres. 7 H7 21
Woodman St. 15
High St. O11 32
Woodman St. 11
Sweet St. H12 29
Woodman Ter. 9 K11 30
Wood Nook Cl. 16 C6 11
Wood Nook Dr. 16 C6 11
Wood Nook Garth, 16
C5 11
Wood Nook Rd. 16 C5 11
Wood Pl. 11
Lodge La. H13 37
Woodside Av. 4 F10 28
Woodside Hill Cl. 18 D7 19
Woodside Pl. 4 F10 28
Woodside Ter. 4 F10 28
Woodside Vw. 4 F10 28
Woodsley Rd. 2 and
3 G10 29
Woodsley Ter. 2
Clarendon Rd. H10 29

Printed for Geographia by Flarepath Printers Ltd, St. Albans, Herts.